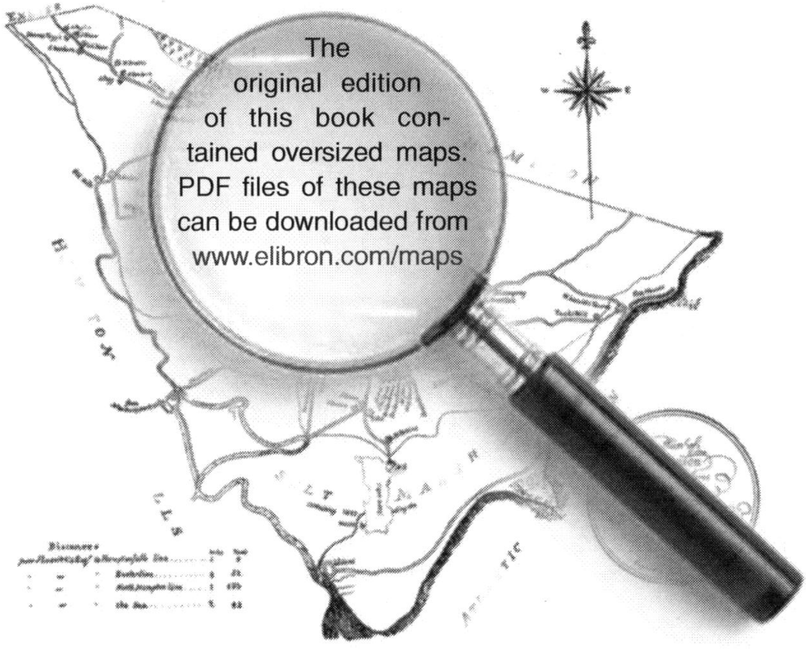
The original edition of this book contained oversized maps. PDF files of these maps can be downloaded from www.elibron.com/maps

AUBYN TREVOR-BATTYE

ICE-BOUND ON KOLGUEV

Elibron Classics
www.elibron.com

Elibron Classics series.

ISBN 1-4021-4518-7 (paperback)
ISBN 1-4212-8965-2 (hardcover)

This Elibron Classics Replica Edition is an unabridged facsimile of the edition published in 1895 by Archibald Constable and Co., Westminster.

ICE-BOUND ON KOLGUEV

J.T. Nettleship

ICE-BOUND ON KOLGUEV

A CHAPTER IN THE EXPLORATION OF ARCTIC EUROPE TO WHICH IS ADDED A RECORD OF THE NATURAL HISTORY OF THE ISLAND

BY

AUBYN TREVOR-BATTYE

F.L.S., F.Z.S., ETC.

MEMBER OF THE BRITISH ORNITHOLOGISTS' UNION

WITH NUMEROUS ILLUSTRATIONS BY

J. T. NETTLESHIP, CHARLES WHYMPER

AND THE AUTHOR

AND THREE MAPS

Westminster

ARCHIBALD CONSTABLE AND COMPANY

PUBLISHERS TO THE INDIA OFFICE

14 PARLIAMENT STREET, S.W.

1895

Edinburgh: T. and A. CONSTABLE, Printers to Her Majesty

TO THE PRESIDENT

OF THE BRITISH ORNITHOLOGISTS' UNION

LORD LILFORD, F.Z.S.

THIS RECORD IS INSCRIBED

AS AN EARNEST OF LOYAL REGARD

Of gentle, strange, unlettered men,
Of Holy Hills, of frosty fen
And ice and island foam;
Of flowers and of antlered herds
And all the nesting of the birds
He brings the story home.

Not his the crystal to divine
With those clear eyes the charmèd page
Where Nature, in the inner shrine,
Enrols her children's heritage.

Yet, when your roses welcome June,
This tale of far-off errantry
May serve to cheat a languid noon
With breezes of the Northern Sea.

PREFACE

MANY like to read of travels in new or curious places who are little interested in scientific results. So it has seemed wise to relegate these to a separate part. For the rest, this is a journal, for the greater part put historically, where the various sides of the life of our island fall into the place where we knew them as part of any day.

To Colonel H. W. Feilden—Arctic Naturalist and unfailing friend to every beginner in the same field—belongs my first thought of going to Kolguev. I feel that I can never be grateful enough to him for lessons out of his great experience and wide research, so generously and untiringly given.

I want also sincerely to thank many other kind friends and distinguished men for that ungrudged advice and help which is so real an encouragement.

Especially my thanks are due to Clements R. Markham, C.B., F.R.S., Professor Alfred Newton, F.R.S., Mr. P. L. Sclater, F.R.S., Dr. Albert Günther, F.R.S., Dr. Edward B. Tylor, F.R.S., Mr. William Carruthers, F.R.S., Dr. Bowdler Sharpe, Mr. Lamont, Mr. Raymond Tucker, Mr. Howard Saunders, Mr. Henry Seebohm, Mr. Harvie-Brown, and Mr. Edmund Meade-Waldo.

The companionship of my friend Mr. Mervyn Powys made the voyage to Kolguev a pleasant one for me. To him is due very much of the success of the venture, for upon his hands fell the preliminary work connected with the sailing. I was as sorry for his sake that Kolguev proved so disappointing a sporting ground as I was delighted to find he had met with some success during his adventurous visit to the wilds of Novaya Zemblya. Credit is also due to him for his subsequent attempts to reach the island. It was not surprising that he failed. The ice borne down with the tide, the constant fog, the treacherous sand-banks, the shallow sea, were odds too great. They would have been serious difficulties even had he possessed—as he could not possess—an intimate knowledge of the dangerous entry to our harbour; a harbour which, because of the bar at its mouth, could never have admitted the *Saxon*.

Later on two ladies, Mrs. Leybourne Popham and Mrs. Ponsonby, started from Vardö in the steam yacht *Blencathra* to our relief, but were driven back by storms. They could never have landed, but I thank them much. I am glad, none the less, that the gallant intention and kind thought were never realised; for 'Rescue by Ladies' in the morning papers would, I think, have been harder to survive than even a winter upon Kolguev.

And one more word.

When it seemed that we were in difficulties, Her Majesty's Foreign Office and the Russian authorities used every effort possible in our behalf. The Royal Geographical Society opened a subscription list, and more friends than I knew I had responded with wonderful kindness. The Committee of the Savage Club were active in making our position known in the proper quarters. The Press, too, most generously gave much room to our case. And in this connection I should like to thank my brother, who, from pleasant country pursuits, was called back to London and plunged into a merciless correspondence on my behalf. Fortunately the event proved these measures unnecessary. But I put these efforts on record not only because I am grateful, but because it is right that I should do so. There is much happiness in the fact that the idea of an Englishman abroad in danger is itself a claim sufficient upon the practical sympathies of all Englishmen at home. Mr. Henry Arthur Cooke, British Vice-Consul, who has proved himself often before a kind friend to any who visits the remote town in which he lives, not only undertook a voyage in the hope of relieving us, but showed us constant kindness during the whole of our stay in Archangel. I cannot thank him sufficiently.

One of the great charms of travelling in Russia is the kindness of the Russian officials. From the highest authority down to the police-master of the railway station, every official with whom I have been brought into relationship has done far more in my interests than I had any right to expect. Particularly I would offer my sincere thanks to his Excellency the Russian Ambassador to London, his Excellency the Governor of the Province of Archangel, to both of whom I owed the papers which carried me through; the Ispravnik of Ust Tsilma, who went even to the length of advancing me a considerable sum of money without further assurance than my word that he would get it back. (Which of us, I wonder, would do this for an unknown foreigner?)

If I have said anything that may seem a little hard of my honest and faithful companion on Kolguev, Thomas Hyland, I beg he will take it in good part. It is not said in an unkind spirit. To take a quiet and inexperienced country man away from his simple occupations to the wild life of an Arctic island is necessarily an experiment of doubtful event. But I trust the experience is not found an unpleasant one now that he comes to look back on it from the quiet of his home.

Searching through old records is always arduous work; and in this connection I cannot but acknowledge the kindness of Mr. J. Scott Keltie, Assistant Secretary to the Royal Geographical Society, and the courteous assistance of every official related to that body.

I am indebted to Mr. Arthur Montefiore, the Secretary of the Jackson-Harmsworth Expedition, for one important reference. Travellers owe much to those who, like Mr. Montefiore, generously give them the benefit of their greater opportunities for literary research.

It would be ungracious for me to close this preface without some reference to the illustrations in this book.

Mr. J. T. Nettleship's beautiful work needs no words from me. Nor does that of Mr. Charles Whymper, unless indeed on this ground, namely, that he has been obliged to limit his artistic power to the simple expression of facts as they were. Also I am under obligations to Mr. Edward Thornton for several careful pen-and-ink drawings from my pencilled or coloured sketches.

About the pictures which have been transferred straight from my sketch-book I need only say that they are correct in detail, as they were done on the spot. For this reason I am sure my critics will forgive an amateur for work that is amateurish.

Often on Kolguev the means of living were found for us by one whom it is difficult to thank; for I refer to old Sailor, the spaniel, one of the most faithful and accomplished dogs who ever followed the gun.

CONTENTS

FULL-PAGE PLATES

The Maps are at the end of the Volume.

INTRODUCTION

As Kolguev Island lies but fifty miles off the coast of Arctic Europe it may fairly be regarded as European. A glance at the map will show that it is the only island in that part of the Arctic Ocean known as Barents Sea.

It seemed time that something more certain should be known about it than could be gathered from chance references in the old books. Also, lying thus between east and west, it held out to a naturalist promise of interesting things.

In the autumn of 1893 I crossed in a trading vessel from this country to the White Sea, with the object of finding out something about Kolguev. I failed. I asked the Governor of the Archangel Province. He was kindness itself, but confessed himself uninformed on the subject. 'But,' he added, 'when my gun-boat goes next year to Novaya Zemblya they might possibly be able to set you down.' It was a kind offer, but the gun-boat would be going altogether too late for my purpose. I asked of the traders and fishermen, I asked of the monks on Solovetsk—but with no result. They knew nothing of Kolguev; only they were agreed that it was a nasty place. So I came back again to shoot the pheasants, no wiser than before.

I knew that it would be useless to go from Archangel, because the White Sea would be blocked till too late for the first summer days. It therefore seemed best to engage a Vardö walrus-sloop to take me over. That was my plan.

In the meantime, however, my friend Mr. Mervyn Powys, glad of the opportunity of sport in those regions, determined to join me. This in every way improved the chances of the enterprise. For he chartered the steam-yacht *Saxon*, of Birkenhead.

She was manned with a crew of Scotch whaling-men belonging to Peterhead, and from that port we sailed on June 3rd. Also we took Thomas Hyland, a young man, who, though he had but lately opened a small business in this country, was glad to come and skin the birds. In all, we were twelve on board.

The *Saxon* is 90 feet 6 inches in length, over all. Her depth is 12 feet 2 inches, with a beam of 18 feet. She is built of wood sheathed with copper. Her draught of water is 12 feet. Her net registered tonnage is 48·58, and by yacht measurement she is 117 tons. She is schooner-rigged.

I do not think that Kolguev is yet exhausted. I think that a naturalist who, fortunate enough to find the conditions of ice more favourable than they were in 1894, would pitch his tent by Lake Promoinœ, and stay there, might do good work.

Whether it is useful or not to have determined geographical points on Kolguev, it is certain that our acquaintance with the Samoyeds is a gain.

Any one who in the future may be landed or wrecked on Kolguev will know that there are natives there, and that they remain there all the year. And he will know exactly where to look for them. He will not make my mistake of expecting to find them near the river mouth, but will move up at once to the head waters of the streams. Here on the Gobista, the Pugrinoy, or the Pesanka for example, he will certainly fall in with the chooms. If he should meet Uano he will be fortunate; if On Tipa, more fortunate still. These are the best men on the island, but all are kind and any one of them would be a friend.

I am often asked about the pronunciation of the word Kolguev. I believe that strictly it is a word of three syllables, but in pronunciation it comes to be a dissyllable, and might be phonetically written 'Kol-gwev. I am aware that a strict transliteration from the Russian would make it 'Kolgueff,' but the 'v' better conveys the sound of the final syllable, and has also the sanction of use.

The Russian name of the curious people who call themselves 'Nyanitz' has been much discussed. That the word is simply *Samo-yedi*, *i.e.* self-eaters, is the common view in Russia. The peasants and small traders of Arctic Russia—men who have lived for generations among the Samoyeds, and are well beyond the reach of books and theories—gave me this explanation as one which admitted of no question. They said that the Russians had 'always' called the Samoyeds by that name 'because they used to be cannibals.' And the Samoyeds, when we asked them, always agreed. The words of On Tipa the Samoyed, when we talked it over in Alexander's hut, were 'Da, da-Pravdah. Tepair nieto, nieto-Davno, davno!' *i.e.* 'Yes, truly. But not now, no, no, long, long ago!' At the same time it would seem not impossible that the idea may have originated from their habit of eating raw flesh. The Russians of the Petchora familiarly address these people as 'Samodine.' This I take to be a diminutive. When speaking of them in the common colloquial way they say 'Sam-yad,' because, as Alexander explained, it is said 'more quickly.' My own imperfect knowledge of the subject does not entitle me to an opinion; so with this I leave the word to the ethnologist.

KOLGUEV AND THE NAVIGATORS

WITH us the name of Kolguev had hitherto been familiar only to students of old Arctic literature, where indeed it fills a very small place. Nor is this strange. A harbourless island with a dangerous coast, it was wisely avoided by any who had a more distant mission on those seas.

1553. *Willoughby's Land* (?)—Sir Hugh Willoughby this year sailed from the Thames to try and find a passage by the North to the treasures of the East (though afterwards he died on the Murman Coast) with three ships, *Bona Esperanza*, *Edward Bonaventure*, and *Bona Confidentia*. The *Edward Bonaventure* was commanded by Chancelor, and Stephen Burrough was among the crew. This ship became separated from her consorts, the other two, after being compelled to return westward, sailing about and trying to make Vardöhuus. And then we read, on August 14th, 'Early in the morning descried land, which land we bare with all, hoising out our boat to discover what land it might be, but the boat could not come to land, the water was so shoale, where was very much ice also, but there was no similitude of habitation, and the land lyeth from Seynam East and by North 160 leagues, being in latitude 72 degrees.'[1]

Where was this 'Willoughby's Land'? Later writers have thought it was Novaya Zemblya, and geographers have put it there. Nordenskiöld believed it to be Kolguev, and this better agrees with the data given, in spite of the difference of 2°.

1556.—And in this year Stephen Burrough, the future chief pilot of England, sighted Kolguev as he went eastward on that eventful voyage during which he discovered and passed the Kara gates.

The account is entitled, 'The navigation and discoverie toward the

[1] Hakluyt, vol. i. p. 236.

river of Ob, made by Master Steven Burrough, master of the Pinnesse called the *Serchthrift*, with divers things worth the noticing passed in the yere 1556.' In this we read on August 25th, 'At a west north-west sunne we sounded and had 29 fadoms blacke sandie oze, and then we were north-east, five leagues from the north-east part of the island Colgoieue.'[1]

1580.—Pet and Jackman sailed this year under commission from the Russian Company to go 'eastwards to the countries and dominions of the mightie prince, the Emperor of Cathay, and in the same unto the cities of Cambalu and Quinsay, or to either of them.' Neither of them of course they reached, but on their return from Waigats they managed to get their ships aground, on August 20th, on the sand-banks to the south of Kolguev.

The account runs: 'At twelve of the clocke we were upon the suddaine in shoale water among great sands, and could find no way out. By sounding and seeking about we came aground, and so did the *William*; but we had no hurt, for the wind was off the shore and the same night it was calme. All night we did our best but we could not have her aflote. These shoals doe lie off Colgoyeue; it is very flat a great way off, and it doth not high aboue 2 or 3 foote water. It floweth north-east and south-west.'[2]

1594.—The famous Dutch explorer, William Barents, on his first voyage 'sayled til the 14th of August, five or six miles south-west, sailing close by the land, which (as they gesse) was the island of Colgoyeu.'[3]

1611.—Richard Finch, in a letter to Sir Thomas Smith, Governor of the 'Worshipfull Companie of English Merchants trading into Russia,' wrote as follows:—'The seventh of August, William and I with our shallop went on shoare. This Colgoieue is a very long and broad Iland with many vallies in it. On the same are many geese, which the Russes use to take with nets in the time of the yere, before they be ouer fledge. In this Iland seemeth to be store of Hawkes. Here William

[1] *Purchas his Pilgrimes*, 1589-99 to 1601.
[2] Hugh Smith in Hakluyt, vol. i. p. 508, ed. 1810.
[3] G. de Veer in Hakluyt. Hakluyt Society's edition, 1850, vol. xiii. p. 35.

Gourdon and our cooper caught two hawkes, whereof one was spoyled in the taking, the other remayneth aliue.'[1]

And William Gourdon himself writes in the same year that 'the third (August) at noone we had sight of Colgoieue Iland and took the latitude, being on the north side of the island, which was 69 degrees 20 minutes: and at night I went on shoare to see the Land, which was high clay ground: and I came where there was an airie of slight falcons, but they did flie all away save one, which I tooke up, and brought aboard. This Ile of Colgoieue is but thirtie leagues from the Barre of Pechora.'[2]

It seems that such Russian coast-maps or charts of Kolguev as existed at the beginning of this century were drawn by the small traders, seamen of the opposite coast. No attempt to fix positions seems to have been made until the years 1823, 1824, nearly three hundred years after Willoughby's death.

1823-24.—In those years Admiral Lütke, during his third and fourth voyages to Novaya Zemblya, crossed to the coast of Kolguev and determined the latitude and longitude of the north-western, and the longitude of the western points, and took views of some points of the north-west coast.[3]

1826.—A Russian expedition was equipped in this year for the survey of the ocean east of the White Sea as far as the mouth of the Petchora, and including Kolguev. This expedition, under the 'under-pilot' Berejnyeh, sailed right round the island in four days, and so executed its commission.[4]

1841.—This year is memorable in the history of Kolguev as being the first during which any landing or attempt at scientific examination was made. Professor Saweljew, accompanied by Dr. Ruprecht (Conservator of the Botanical Museum of the Imperial Academy of Science), paid two visits to the island in the months of July and August of this year. In a most interesting paper,[5] in which Saweljew gives the results

[1] *Purchas his Pilgrimes*, vol. iii. book 3, chap. ix.

[2] *Ibid.*, chap. viii.

[3] *Cf.* Lütke's *Journey to Novaya Zemblya*, Erman's Translation, vol. ii. p. 325.

[4] *Memoires* (Sapiski) *of the Hydrographical Dept.*, v. p. 18.

[5] *Archiv fur Wissenschaftliche Kunde von Russland*, 1852, A. Erman, x. 313-316.

of their voyage, he says: 'We first landed on its southern point at the mouth of the river Waskina, whence Ruprecht made an excursion into the interior with reindeer. After spending ten days here we determined to travel right round the island, and began sailing along the west coast, landing at the small stream Gusina (N. lat. 69° 26') and at the Konkina; however, when we had reached the northern end of the island we were compelled by an unfavourable wind to turn and steer for Swyatoi Nos, on the shore of Timan. In August we again visited Kolgujew, and, since we wished to sail round the east coast, we repaired to Stanavoi Scharok, where we stayed six days, which Ruprecht again utilised for an excursion into the interior. It must, however, be mentioned that our stay on Kolguev was made under the most unfavourable conditions. Of the sixteen days which we passed at the mouth of the Waskina and in Stanavoi Scharok, the weather on ten was such that it was impossible to think of investigations and excursions on the island. The violent storms did not allow us once to leave the cabin.'

Thus Professor Saweljew. The work they did in the six days at their disposal, excellent as it was, could obviously not be exhaustive, and as I spent not six days, but three months, on the island, I was able to add a great deal to it.

In our own day Kolguev has been sighted by the various Arctic voyagers going east and north-east, but I can find no record of any other visit but one.

1858.—In this year a priest, who does not give his name, but who, I believe, was chief priest of Archangel, sailed to Kolguev in a fisherman's boat—probably from Mezen, and has left a curious Russian paper about it,[1] which, in the translation, is very quaint.

'I intended to go to Kolguev,' he said, 'to preach Jesus Christ and to baptize. Before starting there I sang *Te Deums* during three days. . . . The chief thing to be dreaded is the crossing of 150 versts of the open sea.

'. . . I do not desire to speak about all I had to endure during this voyage. One cannot gain much lying on one's back; if you use knowledge and prudence in dealing with the sea you shall never be

[1] Maximow's *Morskoi Sbornik*.

a loser. We coastmen know well how to sail: every child understands how to manage an oar, and every woman—beings inferior to men in every respect—would know how to direct a rudder.' And, further on in the same account:—

'It is true that 130 years ago Barmine, a merchant in Archangel, a Raskolnik,[1] settled in Kolguev at his own expense forty men and forty women who desired to found a hermitage, and all but four died during the first year of their settlement; but they were all old people, and belonged to a very strict sect which allowed during some months to take food only once a week.' This version differs from Saweljew's.

In view of my own experience at the goose-catching, this good priest's account is well worth quoting. He does not himself appear to have witnessed it, but quotes an informant from Mezen.

'It takes us two or three days to arrive to Kolguev; and there birds are come a great multitude and variety—and what a noise they make! The goose gaggles, the eider-duck lets hear its voice, and the drake does not remain behind, and the seagull.

'We begin to shoot it, and during May and June shoot a great deal of it. The dead bird lies in heaps and sometimes begins to stink—but we do not mind it. In the beginning of July the barren geese begins to lose their feather. On the 8th of July a laziness overcomes them. The lazy goose cannot fly,—he has few feathers upon him, and all his down is gone as if some one has plucked him. Such goose sits like an insulted one; he is silent and sad as if hiding himself and ashamed of his nakedness. Now as soon as these geese occupy the small lake, leaving the big one to go there for food, we put our guns aside and took to snares. We spread the snares on every passage from the small lake into the large one for birds to enter in. At the gate we arrange out of the swamp the entrance; sloping towards the lake and very steep in the centre that the goose might not go back after having once gotten into the snare. Having finished this arrangement (which scarcely takes us an hour), we let dogs loose and we ourselves begin to make a noise and to bark in unison with the dogs, showing thus to the geese that they must move from the small lake into the large one.

[1] The Raskolniks or 'Old Believers' seceded from the Greek Church in the days of St. Philip of Solovetsk. They now live chiefly in the villages of the Petchora.

'Now we must be very careful to prevent the leading goose to direct his steps towards the mountain instead of our snares. If he goes there all will follow him, and in that case you will not catch him even with the help of deer. They run very fast. From dogs and men they run into water. We know this and go about carefully, not hurryingly like children. If one goose gets into the gate all will follow him there, and become an easy capture.'

'And if they do not enter your gate, what then?'

'Then we must make him enter; for this we are endowed with reason. But it is also true there are geese and geese, and some of them are exceedingly cunning. They like to hide themselves, and it is not easy to find them out, and even then it is not easy to catch them. Such geese we try to drive into the gate by going in boats to the middle of the small lake, but it is not easy task to do. Geese know men to be their enemies, and are careful. Though of course there are geese and geese. Some are very stupid and easy to be caught, especially the brent goose, who walks of its own accord into our huts. We export from Kolguev about one hundred thousand birds, and could get even more.'

It will be seen later on how excellently this agrees on the whole with the goose-taking as done to-day by the Samoyeds.

This brings to an end all important references to Kolguev Island, which I have been able so far to collect. Very possibly there is more in the Russian, and possibly something still in our own libraries not readily accessible. I shall be very grateful to any informant who may be able to help in this respect.

PART I

Shows how we sailed for Kolguev—Tells of the conditions of the ice of Barent's Sea in 1894—How we at length defeated the ice and reached Kolguev—Of our various landings on that island—Of what it looked like and what we found there, and of how at last, being short of coal and the ice threatening to overwhelm us, the *Saxon* was obliged to hurry away, leaving myself and Thomas Hyland on the lonely isle.

CHAPTER I

PETERHEAD TO BERGEN

June 2nd.—Sometimes those whom chance has taken to Peterhead have little enthusiasm to spare for it as a bleak, uncompromising north Scotch town. But I like it.

I like the fine whaling men who hang about the harbour, with their queer old memories of Davis Straits. I like the whalers themselves, or what is left of them, with the crow's nest and the whaling gear and the quaint figure-heads. I like to lie low and take in all the jangling picture at the fish sales on the quay when the boats come in with the cod. The bay too is pleasant to sail in, with the sea as clear as diamond

ESKIMO FIGURE-HEAD OF THE WHALER 'ENDEAVOUR'

over the barnacled rocks, and the silver sea-trout leaping in the tide.

Also the stupendous steam crane of the new breakwater is wonderful to watch, playing with cubes of tens of tons as though with pebbles; and that other machine —tragically exact—formed of human thews and sinews, and working to the utmost stretch of human capability, as perhaps only convict labour can.

We had time enough for some of this in the intervals of getting the *Saxon* ready for her cruise, though there was necessarily a good deal to be done. But by ten in the morning of this Saturday, June 2nd, all was ready and stowed. With that we weighed anchor and stood out of the harbour, dropping our pilot off the Head.

The wind was light, and this was well; for it was all against us, blowing from the east. None the less the log at sunset showed nearly an average of nine knots and a half. At midnight the wind freshened and went a point or so more northerly, so that we set the sails and made good way.

We found the *Saxon* a remarkably handy little boat which rode the waters like a cork. She was not a comfortable craft just by reason of her very quickness, but this we could forgive.

I had had out the tow-net during the day, but, as it took nothing, I hauled it in again when the water grew rough.

June 3rd.—At ten o'clock on this morning a pair of

THE S.Y. *SAXON*

Face page 4

curlews flying low came circling round the boat. They seemed very reluctant to leave us; several times they stooped at us as if they wanted to settle; they would no doubt have rested on the water, but it was now much broken under a heavy breeze.

Bömmel Mountain—or so we made it out—came into view at a quarter past two that afternoon; and two hours later, about ten miles off the Selb lighthouse, we took a pilot on board. At this time the wind was strong, and there was quite a heavy sea; but the little pilot cutter ran round us like a fish, and the way her two men handled her was a treat to see.

Our pilot told me that they had been out since last Friday, that the Bergen pilots were having a poor time, and that the *Saxon*, whose small size fairly seemed to beat him, was the first yacht over this year.

And so we ran up the big Selbö Fjord, and lay that night at Bergen.

CHAPTER II

BERGEN TO TROMSÖ

June 4th.—If I touch very lightly on these early days of our venture, it is only because so many who read this book will already know all about the coast of Norway—will know its towns and fjords. Why therefore should I tire them by descriptions of places and scenes familiar from the deck of the excursion steamer?

But Bergen was new to me; and as the University Museum was closed for repairs, I crossed the old fish-market bridge and found the little Hanseatic Museum. This relic of the German occupation under the Hansa League is charmingly quaint, and as a revelation of the life and doings of a master merchant at that time, it has a very distinct interest. Dear, dear, what sly old fellows they were! And so complete is this restoration that you want but little fancy to see it all: the traffic in commodities—bought with one set of scales but sold with another (a gain to the merchant either way);—the settlements at the office counter; the lapses into revel when business was over; the poor 'prentices, kept hard at it till late, winking at each other and peeping under their eyebrows as the red-heeled shoes go up the stairs.

There was a good deal to be done in Bergen, for I had purposely left over several things which could be better had in that port. But at half past four in the afternoon we took on our pilot, one Knutsen (and an excellent man, if you should want one), and headed up the fjords.

June 5th.—The next day we had, to begin with, a very pleasant time; for the wind, though northerly, was light, and the day was very clear. We had the tow-net out for a long while, but these waters were strangely unprolific, yielding but a single capture—a polychœte worm.

We unintentionally lost a little way; for Knutsen the pilot took us up the Royde and Vartdals fjords under the impression, as he explained, that we were tourists come to see the sights. So that at noon we found ourselves off Hioring Point, and had to stand out again, passing north-west by Quite Ness, and up the long Harr Fjord. Here we met a strong head wind, the *Saxon* pitching badly and shipping so much water that we had to batten down.

On this evening the sun set at 9.38, but it was light enough all night for reading on deck.

To-day, besides the kittiwakes, lesser blackbacks and common terns, which had been our principal attendants, we came well among the common eiders. On an island off Stadtland (famous for seals) there were a great number of these birds. A very large proportion of them were males, for of course their ducks were sitting. Among them were a few changing plumage, but most were still white on the rump.

June 6th.—Early on the following morning we were passing Hitteren Island, where the red deer are. And—skipping a good deal about lights, positions, and entrances, which would not interest every one—it may be enough to say that in rather more than half-an-hour from the Varnsen Fjord, which we had crossed at 3.40, we came to Rörvik on the Vigtens, which the tourists know.

Knutsen said this was a capital place for salmon; and indeed we saw many nets set and setting—the eiders swimming like tame ducks all about among the people at work. So we slowed the engines while I spun in the current where it set off the rocks. But nothing would tempt them—neither a spoon, nor a fish, nor a spinning-devil. We were obliged to keep the screw just going, and this, though very slow and gentle, possibly scared the fish. I think a salmon is not like a pike in this way; he thinks before he commits himself.

And here besides the eiders, which, as evening came, strung out into long lines and made away, there were many Arctic terns.

I hardly went to sleep that night—it was all so pretty. I sat and sketched it.

Just below Rörvik is that curious lump of rock called

Torghatten. Right through its middle runs a slanting hole which lets the daylight through, as you can see in passing. And beyond, in the distance, were the mountains called the Seven Sisters, and, if you know the legend, you know why. On their eastern faces they were white, but on the west, where the low sunbeams touched them, pink, and purple in the hollows of the

A SUMMER NIGHT ON THE FJORDS

snow. Purple on the left hand also were the long low lines of rock that met the yellow sand. And across the front of this picture came and went the red-sailed Namsos fishing boats, flashing roads of crimson from the rudder to the sun.

At Rörvik itself is the narrowest passage on all the fjords; so narrow that in it two boats of the *Saxon's* size could not have passed at once. And here, in spite of a racing tide, three great timber hulks were being towed, each by a boat with four men for crew. These boats—keks they are called—look, with their high sterns and sternpieces, like little Viking ships. Possibly their shape is a survival; at any rate I liked to think of it as such.

June 7th.—And now we crossed the Arctic Circle, and yet recorded our highest temperature, for it was 70° F. in the sun at 3 P.M.

All day long we were off Norrland, sighting the Lofotens about two o'clock, with puffins on the wing by thousands, and wheeling in the air like rooks. Also twice to-day we saw the great northern diver and several glaucous gulls. We met a large school of grampus, and saw one whale. The men declared it a bottle-nose.

June 8th.—When I came on deck this morning the Vaag's Fjord (68° 40′ N.) was all like glass, and the sunlight was quite dazzling. A great northern diver was flying backwards and forwards from the middle of the fjord to the shore. I saw it do this four times, but could not make out why, for I have never seen a diver catch any fish that it could not eat easily in the water. The bird by its movements should have been nesting, but divers do not nest so early in the year. There was also a large herd of some kind of dolphin feeding round the fjord. Over them hung very many gulls, which seemed (and I watched them very carefully through the glass) to take bits of fish from the creatures' mouths. They would hang above the animal's track, following with keen eyes its course under water, and as soon as ever it neared the surface came down upon its head in a fighting, screaming cloud. The different species of Delphinidæ are always very puzzling to me. I fancy only years of watching in these waters can make you sure. James Smith at the wheel said he knew very well that these were 'Herring Puffers,' and when I asked him how he knew, he answered, 'Because they were smaller than "Herring Hogs,"' and seeing me look doubtful, he smiled pityingly—the conclusion was so clear.

There were also many red-throated divers in this fjord, as in all the rest.

Where the whirlpool is beneath Mount Bentsiortenden, nine miles below Tromsö, we saw our first herd of reindeer coming down across the fjells. Knutsen said that they cross about here, returning in September, though the water is two and a half miles wide, and the current so formidable that, catching the *Saxon* under full steam, it half swung her round.

Just as we were dropping anchor under Tromsö, which we reached at 2.30 P.M. (788 miles in two hours short of four days), I noticed among a lot of common eiders, right across the water, one that somehow seemed distinct from the rest. Presently all these birds rose, and came right along till they crossed the bows of our boat. The odd bird was a grand and lovely drake king eider.

Many of my readers will know Tromsö and its surroundings far better than I, and can skip this next bit, which I simply take straight out of my journal as it stands.

'Tromsö, very differently from all the miserable snow pictures I have seen of it, is now very bright and pretty and smothered in bunting. This year it celebrates its hundredth anniversary. The houses are of wood, the more important with red painted roofs, the smaller roofed with turf, overlying a sheeting of birch-bark. This turf bears a flourishing crop of grasses. The roads are very good, with flagged and well-laid pavements. All along the water-front are warehouses built on piles. To the west of the island of the town rise birch-covered hills. Cross currents here run very strongly; *Saxon* swinging into a new position about every half hour.

'Powys and I crossed with Hyland and the two dogs to the east side, and landed by a stone jetty where is the holding of a most intelligent fisher-farmer, who, as he tells me, has decided to stay at home this summer, after twenty consecutive visits to Spitzbergen.

'We climbed a rocky hill (1400 feet according to a native) covered on its lower slopes with birch, alder and willow, and higher with another species, as it seemed, of willow with very broad leaves. In a bare rock many pairs of ravens had nested, and were now feeding their young.

'We sat for a while at the edge of a plateau which forms the summit of this rock, and then made our way to a cairn on rather higher ground to the east. A golden plover whistled and flew off. Then I noticed about sixty yards off two birds on the ground, which I knew at once to be Richardson's skua, of which I had seen one flying over the mountain a few moments before. Lying down I watched them through the glass. They both belonged to the dark variety, the shade of the back of the neck only a little lighter than the top of the head. One bird was sitting; the other, standing at first, presently settled itself down as if on eggs. They allowed us to approach within fifteen yards before they flew off. I could find on coming up no sign whatever of nest or eggs.

'We walked a bit farther, but beyond tracks of reindeer in the snow, saw nothing of interest till we descended the hill. About half-way down Powys found a titlark's nest with four eggs. Crossing several patches of snow we reached the foot. The dogs put up a pair of rypa, and two others came flying down from above.

'Powys then returned to the yacht. I hunted the lower slopes with Hyland. We found a brambling's nest in the fork of an alder, just as high as I could reach. This nest—made of grasses, goat's hair,[1] wool, feathers of rypa and others (among them the pin feather of a snipe),

[1] Probably reindeer hair.

a bit or two of birch bark (but very little), and covered outside with moss and lichen, and lined with willow-down —contained seven eggs.

'Now we entered a colony of fieldfares. The nests (low down in birch trees), of which I examined several, were in each case entirely composed of grasses and twigs kneaded in with mud; the lining entirely of soft grasses. They were deeper than those of the mistle thrush, and rather less untidy outside. I took clutches of six, five, and four eggs. The birds were shy; none of them mobbed us in the least. On the whole they kept away, only now and then a bird would come and scream, peering through the bushes from about fifteen yards off. Of redwings we saw none.

'On the shore near two houses were the remains of a stranded whale. It was thoroughly well used up. The pelvis formed a seat, the ribs served as rollers for the boat, and the vertebral centra as tackle-blocks.

'Of the many flowers we saw I identified only *Geranium sylvaticum*, a yellow viola (*V. biflora*); dog violet; cloudberry; a ranunculus (*R. acris*), *Caltha palustris*; yellow trollius and a potentilla (*P. verna*).

'Trout and salmon, as they seemed to be, were rising everywhere as we rowed back, and when I turned in at 1 A.M. a band was playing, gaily dressed girls were rowing themselves about, people were fishing everywhere, and there was a picnic party across the fjord.'

June 9th.—The morning after this I paid another visit to the base of the hill and took several specimens of the the only butterfly I saw. This butterfly[1] as it flies looks much like our meadow-brown, but the 'under side of lower wings are grey, like the under side of a sallow leaf.'

I was again 'struck by the fact I noticed last night. Under a considerable proportion of the trees which held a fieldfare's new nest (under the majority I thought) was lying a last year's nest; as if to suggest that the birds return to the same tree and pull the old nest out. For fieldfares' nests are set so deeply and firmly in the fork, that either they could not be blown out in an entire state, or if blown out would surely be carried farther away. I asked our friend the farmer whether boys ever touched the nests, and he said, "Never—why should a boy touch a nest?"

'His little boy Christian brought out a baby hare, apparently about three weeks old, which he had picked up on the mountain side, and was keeping as a pet. This little creature had black-tipped ears, and its coat was pepper and salt. As I held it in my hand it made a squeaking noise.'

While we were at Tromsö we interviewed one Jansen, a walrus hunter, and the only man there, they said, who had ever been on Kolguev. He had landed there once, he told us, five years ago, and he pointed out the spot

[1] *Erebia Manto* Schiff.

on the chart. It corresponded with the Stanavoi Scharok of Saweljew. Indeed he confirmed several points in Saweljew's paper. He spoke, for example, of the abundance of the drift-wood, and said that he found Samoyeds who lived there all round the year, and Russian traders. He knew of no other entry, nor anything of the big gulf.

That evening we left Tromsö.

CHAPTER III

TROMSÖ TO VARDÖ

June 10th.—We did not run in to Hammerfest, but saw it in the distance as we passed; for we went close by the islands Hielmen and Haaien which are its gates. We determined on taking the outside passage as the weather was calm.

Thousands of razor-bills, kittiwakes, and puffins filled the air as we blew our steam whistle off Hielsostauren, and at half-past five in the evening we passed North Cape, and the Nord Kyn at half-past ten. The North Cape, by the way, is actually not quite the 'most northerly point of Europe,' as the books say, for its position is barely over 71° 10′ N., while Knifscher near it is 71° 11′ N.

No doubt it can be very nasty round about this point; but on two former passages I had found it quiet, and this day the sea was almost as still as a duck pond.

June 11th.—At three in the morning we passed a whaling vessel towing a big finner alongside, and six hours later entered the *Ost* Vaagen, as the south harbour of Vardö is wrongly called.

We were all struck by the life and activity of this

harbour. Boats and steamers crowded the place, ponies were being hoisted on to decks, fish cut up and packed, and over all wheeled kittiwakes, herring gulls and lesser black-backs in hundreds.

They do not up here use the words 'keks,' but name their boats by the number of oars they carry; as femböring (5-oared), ottring (8-oared, though now they have but four oars), sekring, and so on. The boats are of the same light and graceful cut as those lower down. The gunwale stroke—splayed out so as to act as a bilge chock—is always painted some bright colour, red, yellow, blue or green, the gunwale itself being invariably white. The rest of the boat is of simple varnished pine. The femböring are the largest boats, and carry besides the 'raaseil' (square sail) a fore and aft sail, the 'sne-seil,' and have in the stern a cabin roofed with pine or birch bark. At the present time there are a thousand of these boats in Vardö.

Vardö also has one hundred and twenty boats of a larger size called 'Kobrumsbots'—boats containing tanks for 'kobs' or seals.

An incredible number of codfish are taken in these waters. Indeed, Vardö is chiefly a huge codfish drying ground. There are miles of split fish drying on rails. The current price this summer of a fresh cod, without head or liver, was 10 öre, *i.e.* one penny. A year or two ago it was worth twice as much.

Mr. Carl Holmboe, the British Vice-Consul, very

IN VARDÖ HARBOUR

Face page 18

kindly introduced us to the merchant who has the rights over the Island of Hornö, and he rowed us over to see it.

This small island rises suddenly from the water, and is formed of mountain limestone seamed with quartz, on the ledges of which the sea-birds nest.

Large numbers of eiders were swimming, flying, diving, and sitting on the rocks. After watching these birds for a long time in their awkward scrambling over the weed-covered rocks, one wondered how it came about that their feet and general build were not better adapted to their habits, for they slipped and tumbled about in the clumsiest manner.

In vain I looked for the beautiful king eider. Whether they breed here or not I cannot say. Mr. Holmboe told me he did not know the bird.

How wonderfully close the common eider sits. I photographed a duck at the distance of two paces. There were evidences, in scattered down and broken eggs, of a good deal of robbery on the part of the gulls. Of course an eider, when of her own accord she leaves her nest, covers up her eggs from sight. And this practice seems sufficient to defeat the gulls; though why a keen-eyed gull should not notice that which is patent to every one else is hard to understand. But, if a sitting duck is startled, away she goes without more ado, often scattering an egg or two over the edge of the nest. And this is the robber's chance.

Kittiwakes we found by far the commonest birds of Hornö. They were nesting in colonies of many hundreds. A few had already hatched, and I picked up a little one dead.

Next in number came, I thought, herring gulls and razor-bills, though puffins ran these close. We counted five nests of the common cormorant, and three of the shag —and these had young, but the others only eggs.

Hornö was a very good place for observing birds. We came to a corner and peeped over, and there below us was such a round of nursery goings-on. For we took the cliff at an angle, so that the whole thing lay open like a book.

In and out of the holes, not six feet from our faces, the consequential big-nosed puffins ran; while herring gulls and kittiwakes were there in plenty with guillemots (both black and Brünnich's) and razor-bills sitting up on end. We were rather a careless party; but could I have crept up there by myself, I should have had a fine time indeed. Nothing would have seen me except a soaring gull straight overhead, for I would have lain as flat as a puft-adder, my very nose hidden; only just two eyes spying through a tuft of grass.

Upon Hornö also I saw three pairs of white wagtails, an odd raven, the meadow, rock, and red-throated pipits, and the snow-bunting. Very handsome the cock snow-buntings looked in their black-and-white dress, and they were singing sweetly. The island held several pairs. We saw the first on the turf roof of a hut by the sea. In vain we hunted for the eggs, and I began to feel it was no wonder that the old bushman had failed to find them, though he knew the birds so well, at Quickjock on the fields.

Once Hyland called to me that he had found a nest, and that the bird was on. But, putting his hand into a deep hole in the bank, he took out, not a sitting bird, but a very forward young one, who had been hiding there. This was a revelation, and afterwards we found two lots of strong-flying young birds among the shingle by the sea.

June 12th.—The next day I visited the other island, Renö, by myself. When I called on Captain Lonnevig (the most northerly Commandant in Europe) he was very kind about this. He is the lessee of Renö, which belongs to the town, and gave me leave to go and see it. I liked Captain Lonnevig very much, he was so frank and genial; and I much appreciated the privilege of being allowed to prowl about Renö; for he is quite alive to the great interest of the island under his care. Indeed we had much in common that way. Only I wished I had been better up in the local flora, for he took the greatest interest in the flowers of Renö, which he said were curiously different from those of the mainland. By the way, Captain Lonnevig assured me that the king eider only came to Vardö in the autumn.

I landed by myself in Renö, in a little bay where is a turfed hut of the keeper—the only person on the island.

Renö was more interesting to me by far than Hornö, because it includes a much greater variety of land. The coast on the south side is most curiously scarped, rising in a series of sloping limestone slabs or terraces. The steps are really from ten to, say, twenty feet in height, and full of crevices, nice convenient places for birds to nest in. Beyond this is rolling grass.

In these crevices nested the black guillemots. I found the sloping terraces slippery and difficult to walk on. But the guillemots didn't. Far from it. They were so bold that they came and settled, and ran into their nesting holes right under my nose, as I lay down and looked over. And they surprised me with their activity; for they lit as lightly as any pigeon on the rocks and ran as nimbly. You could never have guessed from their movements that you were looking at web-footed birds at all. Indeed they appeared to me remarkably like pigeons with extra red legs and beaks. They nested so low down that I could easily put my hand into their nests.

I picked up a little lesser black-backed gull in the down, and brought him home. He was running about in the grass famously—head down and shoulders up, just as a falcon runs.

Eiders were nesting everywhere, in every possible place—the sides of the cliffs, the rocks by the sea, the top of the uplands, and in the marigold bog, side by side with the herring gulls.

For many herring gulls had taken to a bog of marsh marigold, making their nests sometimes on tussocks of grass, sometimes in the middle of a marigold plant. The lesser black-backs, on the other hand, kept almost entirely to the upland grass.

I saw on Renö four Alpine hares, one of them all but white. Also I saw a raven most unmercifully mobbed by the gulls.

The snow lay in deep drifts in all the sudden hollows, and some small mammal had been driving its burrows into the snow.

I wished I had had many hours more to spend upon this charming little island, and I look forward greatly to the day when I can visit Renö again. Not Renö, however, but Kolguev was our destination, and, with the nesting season already advanced, Kolguev would not wait. So this evening, at 10.15 exactly, we weighed anchor and steamed away for the Land of Hope.

CHAPTER IV

VARDÖ TO KOLGUEV

THE Land of Hope was far indeed from being the Land of Promise.

Only one person in all England knew as much about Kolguev as ourselves, and he was now away in a boat too big to take into those shallow seas.

And we knew little enough. The Russians had failed me altogether.

I had been at the pains the year before to go for information all the way to the White Sea coasts, only to find that not a soul among the Pomors could tell me anything at all.

How an ignorance so curious had come about I will explain later in this book. But you can easily see how all this would just give an added element of interest to our venture.

Whether those were right who had maintained that we should find it quite impossible to land at all; whether if we did succeed in landing, we should discover a harbour where the *Saxon* might be secure; what birds, flowers, and mammals we should chance upon; whether we should find people there, or only a desolate and barren land—in a word, the idea of the unknown, this it was which really attracted me, as it has attracted many before.

I do not know that it had a charm as strong for all on board. I think the whalers, when they discussed things in their bunks, very likely condemned it as a poor move which might probably lead to disaster, but could not bring in a barrelful of fat. I think the skipper and the mate both considered it a queer sort of game for a pleasure trip, and very different from the gentlemen's yachting of which they had always heard.

However, I did my best to show them all what we might expect, reading aloud in the cabin after meals extracts from the only two accounts I had been able to find by men who actually had been on Kolguev. I do not know that either of these were very exhilarating. But just see how different they were in those particulars which we wanted most.

For Maximow's priest says this :—'The rivers Krivaya and Gusina are already known to the inhabitants of Mezen as a good resting-place for vessels; and the bay, which cuts deeply into the island, is a safe harbour for the biggest vessels coming from the White Sea: for boats and schooners.'

But this is Saweljew :—

'In its whole circuit Kolguev does not afford a single safe anchorage or road, and its coasts are *extremely dangerous* on account of the slight depth of the sea and the sandbanks or koski.'

June 13th.—This could not be considered an eventful day. Not that any day is really uneventful at sea. At any rate, what with taking temperatures, shaping courses, entering the log, and duties of that kind, we found plenty of occupation. But the entries for this day and the next in my diary contain little more than the log. I give them, none the less, in the hope that they may be useful to some future voyager off this coast.

'A.M. wind SE., strong. Clear, with chequered sky. Shipped a good deal of water. Midnight, wind NE., hazy. Heavy sea from SE.

'Barometer at noon, 29·9, and rising.

'Thermometer, 8 A.M., *max.* 50°, *min.* 40°. 8 P.M., *max.* 50°, *min.* 42°.

'Surface-water temperature, 8 A.M., 39°; 8 P.M., 40°.

'Latitude at noon (by observation), 69° 32′ N. Longitude (by chart), 33° 42′ E.

At 2.30 P.M. were going three-quarter speed, with Kigten Island abeam.

'At 3.45 P.M., because of the heavy sea, we shifted our course from SE. by E. to SE. by S., with Monastery Point abeam.

'6 P.M., Cape Tereberskoi abeam.

'11 P.M., off Voroni rocks; distant three miles.

'We have seen to-day as many pomatorrhine as Arctic skuas.

'A common sandpiper flew round us twice to-day when ten miles distant from land.'

Note.—The most trustworthy harbours on this coast (SE. of Varanger) are—(i.) Novaya Zemblya (69° 42′ N., 32° 6′ E), at the north-west end of the Gulf of Motovski. It is land-locked from every wind, and there is a safe course in by mid-channel. (ii.) The east end of Kildin Island—(its NW. point is 69° 24′ N., 34° 5′ E.). (iii.) Inside Oleni Island—(its centre is 69° 4′ N., 36° 24′ E.). (iv.) The Seven Islands, nearer the White

Sea. (The centre of Karlov, the most western, is 68° 49′ N., 37° 22′ E.).

'*June* 14*th*.—Wind N., fresh. Clear, but with cloudy sky, and appearance of snow falling on the hills at times.

'Barometer at noon 30·1, and rising.

'Thermometer, 8 A.M., *max.* 40°, *min.* 38°. At midnight 37°.

'Surface-water, 8 A.M. 38°; 4 P.M. 37°. Midnight 36°.

'Latitude at noon (by observation). Longitude (by chart).

'At 4 A.M. were on an E. course with strong breeze from N., with heavy sea. Then we changed course to E.S.E., and at 8 A.M. were off Sviatoi Nos. By 8.15 P.M. had crossed the head of the White Sea and had Cape Kanin abeam; distant four miles. From this point we steered E. by S. Log at midnight showed 337 miles.'

This steady fall in temperature seemed to indicate pretty clearly the presence of ice. How far off it was we could not tell: the men thought a long way. At any rate we could see no sign of it at present,—no ice-blink—nothing. And it seemed to me wiser to get some sleep now while there was nothing much to be done. So, telling the watch to be sure and call me if anything happened, at a quarter after midnight I turned in.

June 15*th*.—I had hoped that eight o'clock this morning would see us off the southern point of Kolguev. Very vainly often does man, alas! propose.

I scarcely seemed to have been asleep when the slowing of the engines woke up both Powys and myself. Rushing up on deck he reported ice. The following, we heard, had happened.

It had remained clear till 1 A.M.; then at 1.30 A.M. we had run into a fog, and ten minutes later had met the ice about longitude 45° E. with the Kanin coast distant ten miles. The plan at the end of this book will explain far better than any words exactly how the ice lay.

The skipper was below: it was the mate's watch. At first he stood in for a bit between the ice and the land, hoping for a way through. But, finding that the ice was packed hard against the land, he had been compelled to beat a retreat; and the *Saxon*, now under less than half steam, was feeling her way north-west along the edge of the pack.

Before I go any further, it may be as well to say something about the nature of this ice; for after we had once left it we saw nothing of exactly the same character again.

This ice, then, taken as a whole, was flat—almost tabular. There were of course odd pieces and aggregations which were considerably thicker, but the general run of it was, say, some foot and a half higher than the water. The colour varied from pure white to a peculiar green, with here and there a darker dirty piece, which had evidently grounded somewhere. The green colour is very characteristic of White Sea ice, and I have little doubt that this, the ice which first obstructed us, had come from the White Sea.

There was a certain amount of movement noticeable here and there away out over the pack, but on the whole,

ICE OFF THE KANIN COAST

Face page 28

far as one could see from the bridge, the blocks were pretty much welded together.

But round the edge of the pack itself all was movement. For the ice here was of course quite loose in character, and the floating blocks were spinning round as they struck one another, forming and re-forming little patches of clear water which closed and opened ceaselessly.

Every Arctic seaman who has spoken to me of this since, has remarked that the ice was unusually late in those seas this year. And it certainly was rather a facer to have met with it so soon.

A Hammerfest owner of walrus sloops had told me the year before that I should have nothing whatever to do with ice at this time of year. And of course our little craft was in no way fitted for such an encounter. Had it come to a choice of forcing a passage or of retreat we must have turned tail. But fortunately it never came to that.

When we first came up on deck we were steering north-east, with the Kanin coast on our starboard quarter, having just turned the most westerly point of this first ice at 68° 34′ N., 44° 32′ E.

After this we had rather an in-and-out bit of work as we followed the ice round. Sometimes it led us north-east, sometimes due north, and then again we would have to take a north-westerly course.

At length we reached (as near as we could make it) latitude 68° 57′, and here we found a wide open bight, running about east by north, up which we steamed for some eleven miles. At 8 A.M. this lead came unfortunately to an end, and we had to go north-west once more.

At one time indeed it looked very much as if, with all our trouble gone for nothing, we should have to put about and run back for our first entry. By good chance, however, things turned out better than that.

After a while we found ourselves opposed only by a thin line of very loose and open ice. This was the tail end—so to say—of the big ice-field on our port side, which draggled out as the main body went floating to the west. How far north-west this pack extended we could not even guess. But across the open water to the north of us the ice-blink was steadily shining; so we knew that not very far ahead the enemy would be found again. But we were not fated to encounter it.

Among the links of loose ice there were many pieces quite big enough to damage us very badly if we had met them fair; but by slowing down the engines until we had but just enough pressure on to keep us creeping, we were able to worm our way through—stopping the engines altogether now and then, or reversing as it became necessary either to try and negative the impact of a block or to retreat from it altogether.

At noon it was glorious sunshine, and we were able to get good sights. This north-westerly point of the eastern pack which we had rounded was as near as possible in 69° 16′ N.

Now we crept up along the edge of the ice, going north-east and making from three to six knots. But at last we were able to change our course to the east, and then, going at eight or nine knots speed, we bore down to the south-east.

It was two o'clock in the afternoon when we first sighted long low fog clouds, evidently lying above some land. It could be no other than Kolguev Island. Then I for one felt very happy; it seemed as if we were doing pretty well.

The men, of course, as is always their way, kept saying they could see cliffs, mountains, and so on. But I think it was all but six o'clock before we really made out land that we could swear to, and even then it was constantly obscured, as the fog lay down about it.

And now ducks became more frequent. Flocks of king eider were dotted about, and elegant long-tailed ducks in large lots were incessantly diving close to the boat. Few birds are more strikingly beautiful or more graceful in the water than the long-tailed duck. The drake when in full plumage has a white head and neck, and long white shoulder feathers; but early in the summer as it was, we saw out of many hundreds of these birds only one or two drakes which still had their fine colouring, though they kept their long tail feathers till quite late on in the year. However, as I shall have something to say of these and other birds later on, I will keep now to the voyage itself.

I naturally felt very anxious about our first approach to this island of ill-fame; knowing that the sea about it was, for a very long way out, exceedingly shoal. It might be that we should find the ice grounded and lying all about the island; or possibly in the shallow sea the surf might make it impossible for our small boats to take us in.

The skipper was, I considered, an excellent navigator. But he was perhaps not a thoroughly self-reliant man; and after I found that whenever things came to a pinch he was in the habit of looking to me for advice, I never, at these junctures, left the bridge. And after all, when it really is a case of dealing with the unknown and with conditions which are changing every minute, it all comes to a question of alertness and common sense. In other words, there is a good deal which is instinctive in this kind of work.

Well, at any rate, by seven o'clock we were well within sight of Kolguev; going very cautiously with the lead constantly at work, in some thirty fathoms of water; and so far meeting no sandbanks.

Drawing up in this way we arrived by 8 P.M. some three miles off the coast, and had about seventeen fathoms of water. Then we turned a bit and steamed slowly down.

It was certainly about as miserable and uninviting a coast as you can well imagine. Trees you cannot expect to find in these latitudes, but often their absence is more than made up for by beauty of scenery in other ways—in splendour of glacier or strength of the bastion cliffs across which the sea-birds go in myriads like driven snow. Here we had not this.

We had only a long low line of level monotony; of snow that hung in festoons from the edge of a cliff which seemed not more than some thirty feet high; or again of snow that formed tracks down the cliff's sides and across the beach to meet the ice-fringe of the sea, or opened at intervals to show a wall of sand or clay, washed into darker streaks by meltings and slippings from above. Such was Kolguev as we saw it first.

Now you will remember, from the earlier part of this story, that we had but few data to guide us in our attempt to find a harbour for the *Saxon*. But before leaving England we had quite settled that the Waskina river must be our first objective, and so far had had no reason to change our views. It must either be the Waskina, we thought, or Stanavoi Scharok, *i.e.* Scharok harbour, on the eastern side of the island.

You may very naturally wonder why we should have pitched upon this particular point.

Well, in the first place, in the only days of which we had any accounts, traffic with Kolguev was from Mezen and Indiga. The point at which these traders would naturally touch would be the Waskina, and if there were an entry there much used, one might expect to find it marked in some way by beacons. And the fact that the only line of soundings shown on any chart was just here, viz., from Indiga to the Waskina, pointed to the same conclusion. Moreover, I remembered to have seen in an old coast-line map of Kolguev the word 'Hutte' on the right bank of the river. I also believed that the big lake Promoinœ was the scene of the goose hunts, and therefore on the whole it really seemed that here we might find either a harbour, or at least people who would come off to us in boats.

As we steamed slowly down the coast, we saw now and then a little heap, which the glasses showed to be drift-wood piled. Once too we passed a burial ground, as it seemed to us, with posts and Russian crosses.

I scanned all the coast most anxiously for any sign of a big river, but saw none. Now and again we passed a little river entrance, or what looked like it, but it was always completely blocked with snow or ice.

At last, to the disgust and disappointment of us all, we again met the ice.

It was ice of a totally different character from that which we had left behind. It was hummocky.

That one fact expressed much. It meant that the ice had not, like the other, been formed in these seas. It meant that it was polar ice, come down from the east and north. It meant—though this we did not know at the time—that, driven in upon the shallows by wind and tide, it would form a barrier round the island which none might pass for many a day to come.

And under these agencies much of it was already stranded and piled up on those very sand-banks which we had hoped to round. We could not dare, with the little *Saxon*, to go very close to the pack, for advanced outposts, in the shape of floating masses, any two of which could have sent us to the bottom cracked up like a hazel nut, challenged our approach. There was nothing for it but to turn tail and find an anchorage elsewhere. But where? That was just the question.

We steamed slowly up with the coast about two and a half miles distant. Here we had about seventeen fathoms of water, though we occasionally passed over a bank where it was reduced to twelve fathoms.

At this time we all thought that it was only a question of patience and the ice would shift, allowing us to reach the Waskina. 'Where shall we anchor?' asked the skipper.

Some distance to the north a headland seemed to stand out. 'We will anchor there,' I said; 'we shall probably find deeper water, and shall be a bit protected from the wind.'

The skipper was for drawing straight up to this position, but I did not at all like the chances of sand-banks, so preferred to take the boat out again, and then stand straight in when abreast of the point. This we did, and as we drew in very gradually, the men at the lead giving us twenty fathoms, seventeen, fifteen, ten, we found ourselves at about a mile and a half from the shore in five fathoms of water. Here we anchored. For the *Saxon* drew twelve feet of water, and we had to reckon on a slack tide since the water was now at flood. So we had out thirty fathom of chain, and presently the *Saxon* was swinging with the tide.

While this was going on, there was some excitement on deck owing to the sudden appearance of a seal. And, looking over the bridge rail, I saw my companion come out of the cabin with his little rifle all ready. He shot admirably, but a seal doesn't give you much of a chance when it comes up in that way, for it shows only its head, and it is very hard to judge distances effectively across the water. It is much as if you were to try and hit a floating plum-pudding as it rose and fell at an unknown distance on the water.

That which had looked like an important headland had gradually lessened in height as we approached, until it now seemed scarcely a headland at all, but only a very slight prominence in the cliff, which rose to a height of not more than forty feet.

This was one of our first experiences of an illusion which was to be a great source of distraction to us from that moment—the mirage.

It was 11 P.M.; we had had a pretty long and rather an anxious day, and I think none of us were sorry to turn in. But first of all we served grog to the crew all round to celebrate our arrival. For we really had arrived. Though disappointed about the Waskina, we had so far done fairly well. Ice or no ice, at any rate we were at anchor off Kolguev.

CHAPTER V

OUR FIRST LANDINGS

June 16*th*.—The weather was so still and quiet that it seemed quite safe to leave the *Saxon* for a bit and see something of the island, before going south again to find whether the ice had moved.

So at 10.30 on this morning we got into the small boat, Powys with Duchess his retriever, rifle and gun, I with the old spaniel Sailor, my gun with various shot, and four rounds of ball cartridge. We took with us in the boat Thomas Hyland (our bird-skinner), the skipper, the mate, and four men to pull.

We found the beach partly covered with a coating of ice, formed no doubt by the washing of the waves. This ice stretched away out over the sea, forming a solid platform indented with little creeks and bays. Into one of these we ran the boat, and jumped out at last on to the ice.

Powys took the mate and two of the men, while Thomas Hyland came with me. The other two men returned to the *Saxon* with the skipper to prosecute soundings. It was of course arranged that if anything should call for our sudden return the steam whistle should be sounded.

We had landed just below a big stream, which from our position could be no other than the Kriva River. I was naturally anxious to have a good look at this, which was one of the points visited by Saweljew. And therefore, while my companion went straight inland, I walked up the coast. But the tide was out. We found only a wide river-mouth all sand, not a rock and scarcely a stone visible, with just a stream of water in the channel, three

CROSS JUST ABOVE THE KRIVA
(*From a Photograph by M. W. Powys*)

feet deep or so. On the river sand was the track of a fox. Then we turned south, going gradually inland.

I had some hopes of reaching the big lake Promoinœ indicated on the chart; which would be about nine miles from our assumed position. But of course we travelled very slowly as we were hunting about, and it was not until 8 P.M. that we at last came in sight of a huge lake.

We saw it from a little rise, and it could be no other than Promoinœ; but what puzzled me was this:—east of it by the chart should be a second large lake between the sandhills and the sea. The position of this we could see well enough. We could see the ice lying grounded all round the outer bank, but inside it nothing but a wide stretch of mud. I found out the reason of this later on. At present I simply chronicle the fact.

We had not very long to look, for I felt the wind change on my face and was afraid that it might drive the ice up to the vessel. So we turned and hurried back. After we had been going for some time we heard the vessel's steam whistle, and saw that she was coaling up.

The tide had risen in the river, as I had feared, and ill would it have fared with us had we been a quarter of an hour later. We just did it and that was all; for the hummocky ice was close at the *Saxon's* screw; pushing her up, up, up, until now she was all but abreast of the river. However, they just had time to fetch us off in the sailing dinghy before the river-mouth was passed. Doubtless we could have swum through the river at a pinch, but no one cares to do this when water is almost down to freezing point. This, then, was the end of our first day on Kolguev.

But we ought not to be at the end as yet. It was the big lake that led me into this trap. In order to describe this day more particularly we must go back a bit. And first I will try and give you a clear idea of the country as we first saw it.

The cliffs, as I have said before, where we could see them for the snow, showed north and south much the same character. They were pretty uniform in height, and of sand, or sandy clay. The beach was the same, with ridges of larger stones of the size of cobbles, and bigger, which, with the exception of a few quartz, seemed to be of a slaty nature.

Those cliffs, under the action of frost and waves, were everywhere falling in; splitting perpendicularly in great rifts, which in many places might be traced away on the flat above, twenty yards or more from the cliff's edge.

On all this south-western part of the island we could not find a single hill; the ground was either dead flat or else inclined a few feet to a sandy or peaty rise, or dropped to a moss or shallow lakelet.

But far away, on our left hand, we could see, immensely magnified by the mirage, the faint risings of those hills or mountains of which Bolvana Mountain forms one of the most southern spurs.

From the Kriva's mouth the whole way down the coast until the land drops to the mudflats, runs a line of hard dry yellow sand, which tails off to the east into bog, moss, and grass.

About five miles down the coast and a mile and a half inland we came upon a fair-sized lake, whose shallow sides were full of last year's dead vegetation. We thought we might as well name it, as it was the first of any size we had seen, so we called it Saxon Lake.

Saxon Lake was a wonderful centre of bird-life. Little stints were chasing one another round and round and all about in parties varying in number from a single pair to perhaps nine or a dozen together. I shall devote later on in this book a separate chapter to the birds of Kolguev.

This book will doubtless fall into the hands of many a better ornithologist than myself. But some of my readers who have not had much time or inclination to follow the birds, may like, I think, to have a tip or two, which my ornithologist readers will skip, about each bird as we come to it. This, I hope, will make the book more generally interesting.

The great charm of Arctic natural history lies in its alliances with our own. Go right away into the tropics and you go to another world—wonderful, dazzling but strange. Those birds are here in our aviaries, those flowers we have under glass. But even high up in the circumpolar area you have many of our own old friends among the flowers, dwarfed maybe, stunted by the cold, but still the same; or else just little cousins not very far removed.

And so too in the Arctic, north of Europe, nearly all are birds we call British. By this we mean only that they have all been recorded here at some time or other. Many, as the Tromsö fieldfare, stay with us all the winter, and then go away to nest. Some only stop with us a little while in spring and autumn—just in passing through to higher lands.

And the little stint is one of these. You may see

them running about the tidal creeks till quite the beginning of June and later, and then off they go, returning with their *young ones* sometimes by the end of July. Quick work that, when you consider where they have been.

And not very long ago no one could tell you where they nested. Mr. Seebohm and Mr. Harvie-Brown found out first, and since that time their eggs have been taken nearer home. But never in England. So naturally I looked at this little bird with much interest.

And we saw many dunlins. The dunlin is another of our waders, but it nests with us. They were flying up into the air and making a noise, as they slantingly descended like big grasshoppers. This is one of their courtship practices, and may be compared to the drumming of snipe.

The turnstone, too, was here in his lovely nesting plumage and his orange-red legs. We do not often see him in such fine feathers in this country. A large proportion of those we have here in late summer or early autumn are young birds with more sober colouring, and the old birds then are fading too. For the turnstone does not nest with us. It just gives us a look in when passing north in May, and then later on its return south. This lovely bird has a far more elaborate song than that of any other wader I know. You really may call it a song—I put it down at the time as 'Chĕwāh, chĕwah,

THE SWAN'S NEST

Face page 43

chĕwēēkĭ, kĭ-kĭ-kĭ kee kee'—and he sings it *con amore*, from any little mound.

And we came on a fine playing place of the ruff. In the Cromwell Road Museum is a glass case in which the scene is capitally represented—for a drama of silent life. I think it is really one of the most astonishing performances, more striking by far than the blackcock's lekking, because the bird is so small and so grotesque in appearance. And though they are reducible to some dozen types or so, there are never two exactly alike. However, we have done our best to drive them from their homes in England, and have succeeded far too well.

We saw many pairs of grey plover, and much to my surprise, either none of golden, or but one pair about which I was doubtful.

A pair of large waders flew wildly off. I thought them godwits, but was not sure.

Long-tailed ducks were exceedingly abundant—every little pool holding its pair. A duck brought in by Hyland contained a fully formed egg.

King eider, too, were numerous—I shot a pair. The duck had not been sitting. On one of the lakes I saw one pair of scaup, which rose and flew off, the drake croaking like a crow. Also that day we had the passage with the swans, which, if I detail it according to my purpose at the end of the book, will there be found by those who get so far. And the grouse we found also, which will be the subject of some comment later on.

Besides these birds we found the red-necked and the grey phalarope, and in their stomachs were the remains of mosquito larvæ. So I knew we should be troubled with this enemy when the sun grew hot.

Of flowers we came at once on *Nardosmia frigida*, a well-known Arctic plant, and saw the dead remains of a *Potentilla* and of a large umbellifer. The cloudberry which we left in flower at Tromsö only showed a small green shoot at long intervals. The cranberry was just pushing a few green leaves through the moss. In Saxon Lake I saw what I took to be the dead remains of a *Potamogeton*, and I looked in vain for the *Hippuris* which was so abundant at Tromsö. I made notes on many other plants, but as none of them have popular names, or I do not know these where they exist, I shall be better advised to leave them all to my chapter on flowers.

THE CLOUDBERRY.

We saw no human beings, nor any very recent signs of their presence. But we did come upon old sleigh and reindeer tracks, also upon beacons built up of drift-wood (which was very abundant), and circular furrows round rises where chooms had evidently stood.

Also I pointed out to Hyland a place where stones

had been pulled out of the ground evidently by human hands. Powys picked up two bears' skulls, and a curiously shaped and drilled bit of bone, which the mate pronounced a bit of walrus tusk shaped for a harpoon handle. But I had no doubt that it was formed of reindeer horn; though as to its significance I could not make any guess. It was, as I knew later, a powder measure. Of which more in its place.

Of mammals we saw evidences of none, except a fox's track, droppings, and kennel in the moss; the footprint of a reindeer, and another like that of a wolf. But this no doubt was a dog's.

The bogs were only soft on the top. Two inches below the surface your foot struck the ice, so that it was much like walking on a London pavement in time of slush.

For some time we steamed slowly on in front of the ice. But with the neap tide it fell again to the south. We seemed to be clear of it for a bit, and began to look out for a good anchorage. At midnight, however, the ice reappeared and chased us up, stretching away on our port bow. But as we found we could just keep it in hand without steam, and as it had become important to save every possible ounce of coal, we dodged along the edge of the pack northward under jib. And so we went all that morning.

June 17.—At noon we had the trawl out, and worked

it for about five hours. At the end of which time it seemed better to haul it up as we had taken nothing but a medusa and a slaty stone.

We found the boat sailed very fairly well a point off the wind, and this day we had used the screw very little, just keeping the fires banked against any sudden call.

At eight o'clock on this evening we came to our anchor in $4\frac{3}{4}$ fathoms of water, with 24 fathoms of our port cable out. We were now just off the Sauchika river, as we judged by a dead reckoning. That is to say, we were lying about a mile from the coast, about sixteen miles to the north of parallel 69°.

Powys and I soon felt like clean run salmon, for getting all the lumber out of the bath-room we had a bath for the first time since leaving England. One of the draw-backs to the extreme activity of the yacht *Saxon* was, that even in an ordinary quiet swell the bath, when filled, would respond at once by emptying itself all over the floor. We had therefore made it into a store place for bottles and many other things.

An hour later we went ashore in the small boat. We landed and got off again not without considerable difficulty, because of the surf and the steepness of the beach. Powys walked inland with the mate for sporting purposes. I was alone, except for old Sailor the spaniel, and wanted to inquire more carefully into this approach.

I found that there was no river at all. It is quite true

that from the sea it looks like a river entry; and Saweljew, who no doubt only saw it from the sea, was very naturally deceived; or, if not Saweljew, the navigator who named the place.

There were two small entries, the southern of which was quite closed by a bar, and the northern almost so, except that it was crossed by a little stream which ran out of a lagoon which lay behind the cliff. This lagoon was filled with brackish water. Into it ran a small snow stream across which I stepped. In it I found a lump of red granite and a small half nodule of iron. The cliffs at this point were higher than we found them to the southward, and were overhung with beds of peat.

Just above the lagoon was a sandy rise. On this I found a small sleigh. It was in pieces, but I easily fitted it together. I thought it was a dog sleigh. But the mate, when I showed it to him on the ship, assured me that he had seen the Eskimos of Smith's Sound use just such another when at their sport. They would, he said, fasten some green thing to it, and then, lying down flat on the belly, push it along before them, so that behind this cover they could approach their quarry. All of which I found to be equally true of the Samoyeds when I came to know these people. My find was, then, a stalking sleigh.

I did not go very far, because I was much interested in some small insects; particularly in a sort of little sandfly with very big eyes. This fly advanced by hopping, so that it was not easy to catch, but by patience I effected it. Also I followed the movements of a minute red active bug, which I believe to have been the same as that we call the money-spinner here at home.

A pair of turnstones had made many sham nests, and one was ready for eggs. A Lapland bunting's nest, though empty, was also quite finished.

A grouse was sitting on twelve eggs, of which I took six.

All the while that I was prowling round this part, a pair of glaucous gulls sat and watched me, and once

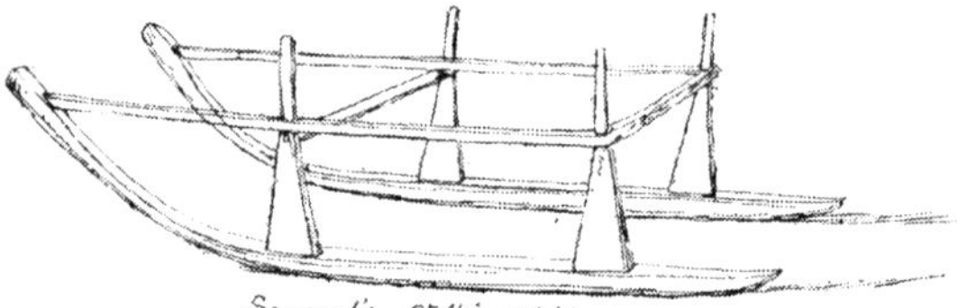

Samoyed's Stalking sleigh.
This is about 3 feet long. To the projecting pieces of the upright branches of willow are tied, and the sleigh is pushed as a screen.

the male bird came and mobbed me slightly. So I thought they had a nest; but search as carefully as ever I could I failed to come upon it.

I chanced on a pool where there was one red-throated diver and a pair of wigeon. A fine yellow-ringed humble-bee came booming round me, but I could not take him, for I had no net.

I had been probing carefully along a line of drift as one who looks for gold, till my back ached with stooping. Straightening myself up I suddenly saw just in front of me a veritable flower-garden—a bank green with grasses and spangled blue and pink. The sun was shining brightly, and the thing was so unexpected and so beautiful after the dreariness that had gone before, that I gave quite involuntarily a shout or cheer, or some noise which made old Sailor come running up.

In a moment I was up the bank on hands and knees, picking and digging up flowers. Sailor at first was much excited at this move, but after he had stuck his nose into several places, and could find no smell of bird or beast, he drew off and sat and watched me for a fool.

Forget-me-not was here, and flowers in purple, white, and pink (*Astragalus*), and a tiny veronica, as well as a buttercup (*Ranunculus acris*).

This was such an attractive spot that we talked of landing again to-morrow. But it was not possible; for we had to shift our anchorage.

June 18th.—Powys had turned in and I was writing my diary in the chart-room, when, chancing to look up from my task, I saw that the sky to the north, which had long looked queer, was growing very curious. Whether it meant wind or fog I did not know, but obviously one or the other was threatening.

I went out on deck. The skipper and the mate were both below, and I routed the watch out of the galley. Bidding him keep a good look-out and report any change, I went on writing.

About one o'clock I was out again. The sky looked

worse. It was clear enough with us, and there was strangely little wind; but across the north of the island an far away to seaward long black and copper streaks of something driving—snow, fog, rain, I could not tell which—made things look very queer.

I did not like it at all. Ours was a most exposed anchorage. The wind was then from the north-west; but it chops so quickly hereabouts that I felt at any time we might be caught. So, instead of turning in, I got out my paint-box and made a careful study of the coast. Immediately opposite the cross on this sketch is Sauchika entry, and from here we could just see the Kriva headland where we lay before.

The most noticeable point in the coast-line from this position is the 'cow's horns.' But I am afraid that a sketch, be it never so careful, would be of little use to a future voyager after a year or two had gone by. Lütke published his projections, and very carefully they were done. Meantime the cliffs have crumbled and tumbled, and all is changed.

As time went on the sky improved, but, all the same, one could not but feel that we had no right to be lying out here while there was a chance of finding a snug place.

The Gusina, if any, promised to be such. Here it was the Raskolniks landed, and here it seemed possible we might too. So, at 4.30 A.M., when the skipper came up, I told him what I thought: whereupon we got up steam and moved off. All looked fair now; there was no particular reason for being awake, so at 5.30 A.M. I turned in.

An hour later they sent down to wake me up. We were just making Gusina Point. I did not much like these tactics. If any sailor should read this, very likely he will laugh at my ideas, for I make no pretence to be a sailor. But I have a feeling that when you get a river debouching like this, and protected by a point which checks the tide, a bank under water is to be dreaded. At any rate there was very little water, it was nearly high tide, and I was afraid of grounding.

So we stood out again, and a bit to the northward, and then came straight for the river which we could see, bearing SE. by S. We felt slowly along, the water getting less and less, and at last came to an anchor in $4\frac{1}{2}$ fathoms, about a mile away from the river-mouth.

It was now half-past ten in the morning. The sea was all muddy from the river water, and we sent

out a boat to sound in every direction. Not a bit of it. Not a ghost of a channel could we find: luck was clearly against us.

The wind shifted to N.NW., and a vile, drying, freezing fog came on: so all day we could not move.

But we made the best of it. Powys kept up our spirits with the banjo, and we sang, skinned, and ate many figs.

June 19th.—The changes in the air temperature in the shade were instructive. When the fogs came on the thermometer fell through 10°, the surface water temperature varying only from 33° to 30° as the ice was near or farther off.

This miserable fog lasted all night, and only cleared at ten o'clock this morning. Then we went ashore to see something of the place and to get fresh water.

We had tough work to get in. Again there was a bar—a bad bar. But over it was running the river very fiercely all in flood, and the water where it met the sea—for the tide was then making—curled and rolled and broke in a very ticklish manner. But we had good men at the oars, and biding our time ran in at last on the crest of a big wave, and found ourselves in a channel no more than six feet deep.

I had taken my collecting gun, so while the man was filling the barrel Powys and I walked about. We had landed on the eastern side. We could see no sign whatever of the old huts of the Raskolniks. Only there were two or three heaps of drift-wood piled up ready for carrying away. The river's mouth was wide and muddy; and all the land about was rolling, lichened, and filled with snow in the hollows. An idea was passing through my mind that we might have to land here for good. And although this was greatly against my hopes you may be sure I looked about pretty critically. But it was a bad camping ground; there seemed not to be a single hollow where a tent could be secure from the winds.

I shot a red-throated pipit, and we saw a pair of twist-tailed skuas.

We were walking slowly along when we distinctly heard a sound like a shot fired. A few minutes more and we heard a second. But after thinking it over we concluded it must have been a cracking of the ice.

It is in the nature of such enterprises as this that the interest, and whatever of excitement may exist, must be rather for those who are doing the things than for those who come thus to them at second-hand. Could I but put my readers fairly into our position; could I but put into them the thoughts and a certain eagerness which inspired us, then I think they would be carried on. Then the dry facts would have a living interest. For see. We had reached, actually *reached* the land which so many had gibed at our hopes of reaching We had now three times been upon it; we had our little craft all safe so far; and on board of her our nets, dredging and collecting gear, our instruments, and all that we could require for a proper diagnosis of this new land. We had everything but the one without which all the rest would go for nothing. In a word, we had no harbour.

Those who can follow this out to the extreme of its bearing will understand me when I say that the situation, dull as it must read, was interesting enough for us.

The tide ran stronger. We waited long about the river-mouth before we could find it wise to venture out, and when at last we pulled out with might and main we had much wary nursing of our little boat before we could take a wave safe enough to carry us over this critical place. We shipped much water, but we got through.

The boat had to go again for water, and coming out was all but swamped. Indeed, seen from on deck, she seemed once to have really gone, so greatly was she buried in a wave which broke upon her. The greatest credit was due to those fine fellows who brought her through.

At noon ice came down to us from the north. It threatened to close us in. There was no help for it; we must steam again.

Off we steamed, and away down the land. It was beautifully clear, with a growing breeze from the north. By 4 P.M. this had strengthened to a stiff wind. We then set sail—our big square sail for the first time—

and, though ice-blink covered nearly all the horizon to the west and south, we saw for a long time no ice. The day wore on—we had passed Sauchika entry and the Gobista mouth. Evening came—we were sailing past the Kriva and still had met no ice. It really did look hopeful. Perhaps, after all, we should at last round the corner and find the Waskina.

Alas! The tide that brought ice on us from the north had certainly pushed the southern pack down, and the wind too had helped to that end. But limit of compression had been reached. We were lower than we had been before: and at 11.30 P.M. were off the sand-bar where 'Shore-Lake' used to lie. To round these sand-banks was impossible. Two miles to the south the old enemy lay in wait; an ice-field grim, uncompromising, impossible. Its better acquaintance we did not desire. We dropped an anchor and stayed.

June 20th.—When I turned in at two o'clock this morning the situation was unchanged. I had waited and waited hoping against hope that the ice would move and let us pass.

I slept too soundly by some bad luck. Usually I was like some weasel always popping a nose out of the hole every hour or two to see that things were right. This time I had clean lost consciousness for nearly five hours straight off, and was only awakened after seven by the sound of the screw.

This had happened. The watch had been badly kept. At half-past six ice had been reported, coming up. The engine had only half pressure (25 lbs.) on. The ice came up with extraordinary rapidity right in the face of the wind. Nearer and nearer it had come until it was close up to the vessel's screw. Mercifully the steam was just up—but only at that moment—and we had slipped out of the dragon's mouth. It had been a narrow squeak—too narrow to be pleasant by a very long way. I must confess I was glad I hadn't been on deck a moment before. If a block of that ice had touched our screw ten chances to one we should have paid dearly for it. Had I been on deck an hour ago we should then have been tacking away. Vain questions do no good. We didn't ask the skipper why he hadn't tried the sails. All the men were very grave and silent. They had all had an object-lesson which they would not soon forget.

We had now, however, a little notion of what this ice really was. It was, as the skipper put it, 'real, solid, polar ice—same as we had in Smith's Sound.' Rough, jagged, tumbled, hummocky, it had come, setting south-westward, from the Novaya Zemblya seas.

As long as I live I shall regret this day because of a fatal mistake. We should have landed at Kriva for good and all; and we didn't. We could have done it, but we lost the chance, and it never came again.

Reluctantly, slowly, we moved up in front of the ice.

It pushed us up, so to say. As long as we had it well in hand we didn't care. We still clung to the hope of a harbour, and harbour there was none on this side. If this ice was coming on, the northern ice might be going back. It seemed that we might yet find a way round by the north when that ice went up with the tide.

We sailed as much as we could, saving our coal all we knew. But I had made up my mind on a point which, before I turned in, I unfolded to my companion.

'If,' said I, 'we can't get round the north of the island, then I land to-morrow.'

Now will I enter upon considerations which I should not deem it necessary to give, were it not that, since my return to this country, I have been taken to task in more than one quarter for the 'folly and rashness' of committing another and myself to such uncertain chances as offered themselves upon the island.

But the reader who reflects that we had the evidence of our eyes that there had been, if not lately, at any rate not very long since, persons on this island with their reindeer and their sleighs; that even if these people had crossed to the mainland they would be coming back again when the sea was clear; or else that some from the mainland would come to them—for nothing is more certain than that even Samoyeds would not long exist without flour and other necessaries of life;—and finally, that we had come all the way with no other object but to inquire into this very country—the reader, I say, who reflecting thus will put himself into my place, will, I venture to believe, acquit me of everything but a very sensible intention to carry out, as I might be able, the object at which we had been at so much pains to arrive.

And I did not dislike this island. 'A good island; may God bless it,' people had said in the priest's paper. And these words had stuck in my head.

But, to be sure, my thoughts were not shared by all. Our old Greenland hands had conceived a very immovable prejudice against this place. They could not imagine what in the world any one could want with such a miserable-looking and inhospitable district. No whales, if we except a beluga who showed for a moment at the Gusina mouth—no bears—nothing that seemed to them worth the trying for.

And this part of the consideration weighed much with my friend, and very naturally, as I could not but admit. For not every one is so much concerned in flowers and birds and objects of nature that he will venture himself for them into the unknown. And to a keen sportsman ever hoping for large game worth the killing it seemed indeed a promise of disappointment. So, very wisely, Powys determined that he would journey still further afield. To Novaya Zemblya he would go, and he went, for bigger quarry.

He was very much set against my purpose. He urged that I might not find these people, and what should I do then, with much to the same effect. I acknowledged the kindness and good friendship of this; but the mind which is completely made up is no longer open to alternatives.

And so ended the sixth day of our disappointment.

June 21st.—The morning settled it. The wind was chiefly north; but sometimes it was north-west, and then in little over an hour's time had swung right across to the north-east. Anyway, it was bad as bad could be for the northern ice.

Over the north of the island swept the same fog-clouds, black and copper, and wicked in look. When they lifted for a moment there was the ice-blink, as steady as it had been before.

We talked over our position. It came to this. We had just enough coal for a run to Vardö—we had thirteen tons or so—but not more. To go on playing this game of dodging and waiting was to ruin all. What would I not have given to be back at the Kriva! But it was now too late; the ice was chasing us up.

And so I had to land.

And then this question arose. Should I go alone or should I take Hyland with me? I was strongly inclined to go alone. I mentioned it to Hyland; he begged me not to leave him behind. And when I pointed out to him that it would be no child's play—that, indeed, we might not come upon the natives at all and should have to trust to our guns for food, he still said he would come. I liked him much for this; because I knew quite well that his friends the sailors, who had formed the gloomiest views of the place, had told him in effect that he would only be going to his death. However, it was settled: Hyland should come too.

We had this understanding. We should take a month's provisions, which might be made to last five weeks. Also one of the little tents, the instruments, and a few other necessary things. On landing we should go into camp, and then, leaving all we could not carry, walk down the coast to the mouth of the Waskina and try for the Samoyeds. Should we fail in meeting them there then we were to travel on to Stanavoi Scharok (*i.e.* Scharok harbour), which we believed to be the place where the Russian gunboat had lain last year for a few hours, according to information which Powys had obtained from the Russian Consul in Vardö, and had in writing from him.

Any plan or change of plan I undertook to put on paper, and to bury six feet due north of a cross which I should erect at such points as we might reach. And the further to protect it from busy hands, I was to write on it these words in Russian, 'Nicholas the Priest,' because it seemed that this, could it be read, would make any aggressor pause. I here calculated on the reverence or the superstition of the native mind.

Then the *Saxon*, after coaling at Vardö, should run for Novaya Zemblya, and returning in a month's time call for us. We should still, we thought, have time for some work at dredging and in other directions, which we could not do without the boats.

But, though we all *thought* the ice would be gone by that time, none of us could pronounce on such a point as this. This consideration added to my friend's anxiety on my account. So then I relieved him of all concern and of all obligation. 'Do not worry. If you can come, come. But if not, then we will cross with the Russians or the Samoyeds, or with any that do cross that sea. Or should the Russian gunboat give us a call we will go with her; and that is all one.'

Because the year before, when I was in Archangel, the governor had been so kind as to offer to send me in the gunboat when she went to Novaya Zemblya. And although that would have been too late for my venture, I had from Vardö the other day sent a note to Archangel inviting the officers to call and see us when they came that way.

So thus it was arranged.

The men knocked up a little sleigh, on which we entertained I scarcely know what hopes of bearing some things across the tundra. And at 8.30 in the evening we got all our packages into the boat and pulled away to Sauchika entry.

We were foiled.

For the wind was against the tide. Great rollers were running up and smashing on the shallow sands, and though we hung about for a bit we could not make it other than an undertaking far too dangerous. We had all but left the return till too late. The ice, as we pulled back, was fairly *racing* up upon the yacht.

How wicked it looked! Monstrous lumps indeed they were, pounding on, one after the other, dipping and swaying like great hungry sea-bears thrown forward as cavalry from the pack.

There was an old man pulling in the stern thwarts who had spent a lifetime in the Greenland seas. He made an impression on me I shall not hurriedly forget. His hard old face, half fear and half defiance, he never stopped inveighing at the ice. 'Damn thee,' he said, 'damn thee, are ye coming on then, ye blackguard, are you going to have us this time? God Almighty, but it's close you are!—Now then, boys,—oh, the devil!' and we just missed a big block. Thus ran his adjuration; only he said it in the Scotch, and he said it in the whaler's form of words.

'You don't seem very fond of the ice, Jim?' I remarked.

'I have lost too many relations by it,' he replied, 'and blood is thicker than water.'

There came to my tongue-tip, but I did not say it: 'And thicker than either is ice.'

We were on board again, catching up the boat and slipping off only just in time.

You can imagine something of our feelings. I

knew well enough there was now but one chance left.

That chance was the Gusina River. It would put the whole island's length between the Waskina and ourselves. But it was that or nothing this time. Now you partly see why passing by the Kriva was so fatal a mistake. We steamed north. Suddenly we were moving in a dead calm: the water was like oil. This was eloquent—a sign we had learnt to understand well. But the ice had moved to the north of the river; so we could land there if the tide allowed.

Once more we filled the boat and pulled off. As I said good-bye to Powys, whom we left upon the ship, I promised I would stay in camp at least till the following evening to assure myself that the *Saxon* was through.

What if the northern and the southern ice stream, formed by the island, came together a few miles to the west?

The skipper said he had no doubts on this score, and most earnestly I trusted he might be right. But who could tell?

We reached the river-mouth. All was well. We pulled in and landed, not on the right or northern bank where we had landed before, but on the left or southern.

I chose the best spot that offered for the tent on the sand of the river-side, under the lee of a bank which separated the river from the sea, and there we stowed the things.

It was just on midnight.

Of all the depressing partings I have ever had to get through this was the worst.

One by one the sailors came and shook hands with us as with doomed men. I rallied the mate on this glum proceeding with 'Nonsense; you wouldn't mind coming, would you?' 'Now, sir, look here,' was the mate's

GUSINA CAMP

reply, 'I wouldn't pass a month on that there island, not for a thousand pounds—there, not if you was to give me the *Saxon*.' Then he shook Hyland by the hand, slowly and lingeringly, and uttering words I could not catch. But Hyland told me what he said.

It was, 'Well, good-bye, good-bye for *ever*. I shall never see *you* again.'

Poor good fellows, they meant it well enough—and I am sure their hearts were right. But I was very angry at the time; it seemed so disproportioned to the very humble venture we had before us. And I thought it so hard on Hyland.

But the men pulled off.

While we were arranging the tent old Sailor was sniffing round, and came to a halt on the top of the bank, his ear cocked, pointing down into a crevice. There, far down, we found a snow-bunting's nest which held six eggs. It was a pretty incident, and I took it for a good omen.

The sun was getting higher, and the river floes and all the banks were flashing wonderfully.

We could not have landed at a more delightful hour; and we were in the best of spirits.

PART II

Shows our life in our first solitary camp — How we followed vain hopes — Of foolish wanderings — A seal hunts for ducks — How we walked right across the island — Our interests by the way — Of snow gullies — Of icy rivers — Of the birds and animals — The geology and natural features — And of how, finally, we fell in with the Samoyeds.

CHAPTER I

IN GUSINA CAMP

June 22nd.—An hour after we had landed the *Saxon* was under way. For a long time I watched her standing away to the westward, and then set to work to get things square.

You saw in the picture how we were just separated from the sea itself by a clay bank, so that we could not work and watch too.

The actual fixing up of the tent did not take us very long, for these Whymper tents are certainly good in that way. But of course we had a good deal to do besides that.

It took us some time to collect wood for the fire. The beach was strewn with drift-wood, and some of it was heavy to bring in. Hyland, who was far better than I at carrying weights, really surprised me by the enormous logs and trees he bore in on his shoulders. Many of these stranded trees had been chipped by the axe on the spot where they lay, and some of them quite recently. Of course we looked at them with great interest, as they were, if we except the sleigh-tracks, the first evidences we had seen of human occupation.

After a bit we had a fine pile of drift-wood collected, and, making a roaring fire, cooked some supper.

Hyland was very anxious to shoot a grouse for breakfast, so I sent him off with his gun. When he returned at 3.30 A.M. he brought with him a fine cock bird, and told me that the *Saxon* was still visible, and going ahead all right.

It was not at all a pleasing morning, for a chill fog had settled down over the island. However, it cleared away about nine o'clock.

We were sitting at breakfast when we distinctly heard, as we both believed, two shots fired. The sound seemed to come from an easterly direction, exactly from the same point whence Powys and I had thought we heard them the other day—last Tuesday.

It at once struck me that our wisest plan by far was without any delay to try and find this mysterious shooter.

Unfortunately we were on the wrong side of the Gusina. However, we took our guns and old Sailor, and followed down between the river and the sea.

And now it was that we had the first experience of those snow-filled gullies which were to give us so much trouble later on. Many of those we met with in the interior of the island were much more formidable than these; but I think I may as well describe them once for all.

Imagine, then, a deep, narrow dene or combe with very suddenly sloping sides. This in the winter would be entirely filled with drifted snow. Now, as this snow begins to thaw, it drops, of course, towards the middle, until it lies at the sides at an angle say of 45°. At the bottom forms a stream, the banks of which vary in height from six to twelve feet or more, and are either perpendicular, or more frequently sloping inwards, so that the edges overhang. You will understand at once what a nasty place this would be.

A GULLY

But very often the stream itself was out of sight—entirely bridged over with snow; only you could hear the water plashing and tinkling somewhere down below. Here, then, was a regular trap. If by any chance one had fallen through the snow-bridge or had otherwise found oneself at the bottom, I really do not know how one could have got out. For we had no ropes, nor any appliances for snow or ice work.

Old Sailor, too, was an anxiety. The old dog was pretty active, considering his years, but I was always afraid of his getting caught. He was very quaint about these places; inspecting them most carefully before he would commit himself in any way. He didn't at all like this mysterious sound of water in the depths, and when he came to a little bit of a thing which I could step

across, he would make the most prodigious leap—landing often with a couple of feet to the good. This satisfactorily accomplished, he would just look over his shoulder at the place, and then rush about like a puppy and roll in the snow, as much as to say, 'That was rather a fine jump—another of them defeated.'

Well, at last, after following several curious horse-shoes which the river makes, we came to a grassy cliff from which there was quite a panorama. Below us the river made almost a circle, ringing round a peninsula of sand and grasses, where was a Russian cross, marking, as it seemed, a burial-place.

A peregrine falcon rose as we approached, and we found the nest half-way down the cliff, just under where I have put the star in the sketch.

The nest, on a little projection in the grass cliff, was a simple depression in the ground, scantily lined (though it was scarcely a lining) with dead grass and a few bits of down, no doubt from the sitting bird.

It contained four eggs.

The cliff at this point was not in the least degree precipitous: it sloped, rather than fell, to the water's edge. But there was an entire absence of castings and of remains of food about the nest. There was not a bone or a feather to show the predatory nature of the occupant. I afterwards found a few pellets about the Russian cross—an obvious resting-place—but by the nest nothing.

My readers will know that all the birds of prey, and many others besides these, throw up or eject the indigestible part of their food—the bones, fur or feathers—in the shape of castings or pellets. You can see the bird 'choke' much as a ruminating cow does, and then the pellet appears. Any one who has kept these birds in captivity—any falconer, for instance, knows quite well that it is no good offering the bird a new meal until this part of the last one has been got rid of. It is very necessary too for the health of these birds that natural food of this sort should be given. When you see—as you sometimes do see—a collection in which the birds of prey are fed on plain horse-flesh or liver, you may feel sure that they will not remain long in good health. Examine these pellets every day to see whether all meat has been properly digested, and you have at once a simple key to the success of the *regimen* and the bird's condition. But this is a digression.

Now, some raptorial birds never let these pellets be seen near the nest, but others are not so particular. Our screech or barn owl leaves its mouse-bone and mouse-fur pellets all about its nest, but our brown or tawny owl either moves them away or is careful to eject them at a distance. At any rate I have never seen any castings at a tawny's nest.

On the other hand, you will find reliable accounts of peregrine's breeding places in which not only pellets were seen but much old food, so that Yarrell says: 'The presence of birds' bones in or around the nest seems to be the rule, and upon the top of the cliffs near St. Abb's Head, where Selby visited a nest, he noticed, scattered in great profusion, the castings of the falcons.'[1] But it is quite obvious that, in such an instance as this, the young birds were hatched; the food was for them, and the castings were theirs.

Just opposite the Russian cross the river, widening

[1] *British Birds*, 4th ed., i. 59.

out, seemed to be shallower. I took off some things and waded across, while Hyland sat with Sailor and watched. The river was about a hundred yards wide, and there were some nasty-looking places where the current set very strongly. So I went slowly up and down, feeling my way along the shallows with my toes, for you couldn't see the bottom, it was so muddy. At its deepest it was only half-way up one's thighs. Snow water is not warm, and for a few minutes after landing I could not tell my feet from my boots.

Hyland came manfully across in half the time that I had taken. But, though I am afraid I chaffed him a bit, I was really very sorry for him, poor fellow; he felt the cold so badly. It seemed to catch him higher up than in my case. I fancy that the practice of having a cold bath every morning perhaps makes one less sensitive.

Sailor thought nothing of it, though he had to swim. Dogs must look on us as very poor and imperfect creatures in view of the fuss we make over a thing so simple to them.

Under the cross I found a wooden implement which I took for a paddle. I thought possibly it had belonged to one of the poor Raskolniks, but I know now that it was a Samoyed dough-trowel or spatula, and that this was therefore a Samoyed grave. All of which will appear later on.

Just after this we came upon a fox-warren. If you

look at the end of Sir George Nares' book on the Arctic Expedition of 1875 you will find a most interesting account by Colonel H. W. Feilden of this animal's habits as observed in Grinnell Land. Had I known of this before I should have made a closer inquiry into the interior of the holes. As it was, I noticed only that the place was strewn with the remains of geese, and that the sternum and fore-legs of a small seal were lying there. I was on the look-out for the remains of the lemmings, so I did of course look down into the burrows as far as I could see; but very possibly a little digging might have revealed the larders described by Colonel Feilden. We could not, however, afford to wait if we were to find the firer of shots.

It was not long before we came upon the track of a sleigh drawn by five reindeer. This was very cheering, for it might have been made yesterday—it looked so fresh.

We began to follow that sleigh-track. It doubled and twisted all about. The man had been collecting willow-grouse eggs, or else the dog (for he had a dog) had been collecting them: I could not quite make out which. But we saw several plundered nests and places where a bird had been shot, or more probably caught by the dog. Often it was very hard to follow the track. I was very much struck with the lightness of the runners. Of reindeer sleighs I at this time knew nothing, of course.

On the lichened peat the traces of the runners were often quite invisible, but this did not matter as the reindeer's feet had always split the crust a little. By far the worst tracking was over the dry moss of the old bogs. This was so wonderfully elastic that we always lost the track on it and had to cast about and pick it up on the other side. Our own steps in the same way left no impression on the dry moss.

Well, we stuck to this work for about three hours, until, in short, the track took us back in a ring almost to the Gusina again, when it suddenly turned and went straight off in an easterly direction. The man had gone home—wherever 'home' was.

And then I came to the conclusion that I didn't think much of following five reindeer and a sleigh; and that home was the best place for us also. I learned afterwards that the mysterious hunter was On Tipa. Had we only succeeded in finding him we should have been spared many days of weary walking. But we had not. So there was nothing more to be said, excepting what I uttered at the time, and that was very short.

At this moment we were on a bit of rising ground, and looking out to sea we saw, apparently in the middle of the ice, a sail. With my glass I made her out to be of a sort of cutter rig. No doubt she was a walrus sloop.

This time we took a slightly different line to our ford. As we came in sight of a little round lake a snowy owl, either a female or a bird of last year, stooped and picked something—a fish I suppose—out of the water. She sat on the bank with the quarry under her foot for a minute or two before skimming off, and a red-throated diver swam backwards and forwards in front of her with its bill wide open, as though daring her to come on again and steal its lawful supper. Doubtless it was making a noise of defiance, but we were too far off to hear. I saw the details of the drama through my glass.

This, the most lovely of all the owls, comes to us at intervals. It has never been proved to have nested in Britain in the wild state. An Arctic bird, I have seen it frequently in Canada, and have described some of its habits elsewhere.[1] This was the first time we had seen it on Kolguev.

To-day also we saw a white-billed northern diver flying past. This bird I never saw again on the island. The white-billed northern diver is quite distinct from our bird. I consider its rarity in Kolguev most remarkable; for it is known to be a breeder in the lakes of the mainland tundra.

In the bank of a little brook we found a nest of the red-throated pipit, which contained six eggs. The nest was built entirely of grasses. No one, I think, could mistake this bird for its British ally, the meadow-pipit.

We moved a grey goose off her nest of four eggs. I believed her to be a bean goose.

[1] *Pictures in Prose: of Nature, Wild Sport, and Humble Life* (Longmans, 1894).

And now, before we close this day, comes something which involves a moral, if one only knew how to point it. We often speculate on the law of chances; we wish, for instance, we could trace the steps, see some of the working by which things which seem as if they must fall out in such and such a manner, are just missed, just fail.

Well, the very first thing that met our eyes as we once more reached the bank of the Gusina at our crossing was the fresh track of nine reindeer and two sleighs all up the side of the river mud. They had passed since we crossed. So fresh was the track that the mud had scarcely settled in the footprints of the deer. There they were, sure enough, a light sleigh drawn by five and a heavy sleigh by four deer. The men had probably been up the river collecting drift-wood. We traced their trail for some way. One more turn of the river bank and they would have come full in sight of our little tent, where we should doubtless have found them on our return. But nothing told them; and they stopped short. How they missed seeing our footprints as they passed I couldn't imagine. Fancy, if you can, an Indian missing them! When I came to know the Samoyeds I found they were not Indians—by a very long way.

But see the case: Here were we, the only two, solitary white men on an Arctic island, in need of but one thing, and that was native help. We knew absolutely nothing of where these natives lived, but this very day we had missed them by an hour or two only, if so much, and missed them twice.

It took us six days to find them after that.

So we had to wade the river again. And while we were undressing old Sailor curled up and went to sleep. We didn't notice it at the time, but when we had finished dressing on the other side he was not with us. And, looking across, there was the old dog on the mud sleeping as sound as a bell.

We had the greatest difficulty in waking him up, and at one time it really looked as if I should have to cross again for him. But at last when I had exhausted every view-holloa and yell in my glottis, and Hyland was voiceless, he quietly raised his head, looked round, rose, shook himself and swam across; coming out as innocently as though he had been waiting all the time for us, and not we for him.

It was just 10 P.M. when we reached camp. All the scene had changed. There, where had been the open bight by which we entered, was a solid pack of hummocky ice. All was ice—ice to the northward, ice to the southward, ice packed tight and grounded against the island, ice to seaward as far as the eye could reach.

My diary for this day closes with the words, '*Saxon*, I trust, is through.'

CHAPTER II

PREPARING TO START

June 23rd.—So there was nothing for it now but to make a determined effort to find the natives. Obviously this was the first thing to be done.

Our trial trip of yesterday had shown me one thing pretty plainly. This :—

If I held to my original idea and should try to reach the Waskina by travelling down the west coast, we should have entries to go round, and at least two big rivers — the Gobista and the Kriva — to cross. Our passage of the Gusina did not altogether tempt one to repeat the process more than was absolutely necessary. Also the sleighs we tracked yesterday had made away for the east.

So it seemed better in every way to steer for Scharok harbour, keeping as much as possible to the high land which would form the divide between the watersheds.

Knowing nothing of the interior of the island, it was impossible to say how long such a walk would take. It might be, too, that we should arrive at Scharok only to find that there were no natives there. What should we do then?

Well, in this event we would, I thought, walk south from Scharok, and keep on walking right round until, if we had found no natives, we should arrive at last back at our Gusina camp.

But, of course, there was a good deal to be done before we could start. Every little detail had to be carefully thought out; nothing left behind which we might want, but nothing taken that was not absolutely necessary. For everything had to be carried on our backs.

Meantime it was pleasant enough in camp. With a light breeze that varied from N. to NE., a barometer that stood steady at 30 in., and a sun which at mid-day showed 62° F., the day was as beautiful as well could be.

We were well occupied with egg-blowing and bird-skinning, so we did not leave camp except to climb the bank and look out over the sea.

Close as the ice was packed, there were yet a few bits of open water on the shallow, just large enough for a paddling-place for a few birds.

I liked to stand on our little cliff and keep an eye on the life of these ice-girt pools.

They were always great playing-grounds for the long-tailed ducks, and I find this about them in my diary:—

'*Harelda* has at times a remarkably human cry. It is funny when one looks down on him from the cliff, to see him steering his way with rapid feet among the

THE RIVAL WILLOW-GROUSE

Face page 83

hummocks, head under his wing the whole time. He never seems to look where he is going, yet never makes a mistake; avoiding in the deftest way the floating lumps of ice. But his mate has the caution of her sex. She swims with head alert, and both eyes open.

'We have not yet succeeded in finding the time when birds are quiet or not moving. It is now 11.45 P.M. (Kolguev time). The sun is just getting up again. Cock willow-grouse are challenging one another, and

OUR LITTLE CLIFF

fanning their tails just across the river. Geese are flying about, waders and small birds singing as they have done all day. However, as while on our march we shall sleep at mid-day and travel all night, we shall soon know more about it.'

Yes, I can see that scene now.

The rising sun had just caught the opposite bank of the Gusina, and a mist was rising from the mud and the ice sheets on the river. Down into the sunlight strutted a splendid old willow-grouse, his wings drooping, his tail fanned. On the very edge of the bank he stood and crowed, bowing like a Mussulman till his head touched the ground. He did more. He went through a kind of dance, turning round and round, and stopping now and then to flirt his tail, and answer with all the power of his lungs his rival on a little hillock not twenty yards away.

June 24th.—I was writing in the tent this morning at 1.30 when Hyland, who had been walking along the shore, came in with a very interesting piece of news.

Some creature, a seal he thought, was hunting ducks among the floes.

In a moment I was round the corner of the cliff and at the water's edge. It was true enough.

A group of long-tailed ducks were just settling on a little bit of open water close in front of us. As they lit they drew up together in a bunch. Before long there was a panic among them, and they rose wildly in different directions. Right in their midst—or what had been their midst—appeared the dark head of a seal. Risen from below, had he been trying to take a duck, or had he not? The point was soon settled.

The ducks flew round and lit again in the next pool. The seal raised his head a moment higher from the water, and then sinking, disappeared. We watched the

A HUNTER OF DUCKS

Face page 85

ducks. Once more they rose confusedly, and once more the black head showed.

Again the ducks settled down on the water, just on the other side of a small flattish floe. We moved along till we were opposite this, and once more watched. And now close to the edge of the floe a seal's head twice appeared. The creature raised itself high above the water and looked about, reminding one exactly of a weasel sniffing the air for a mouse when the hunt is momentarily checked. Again the head disappeared.

Half a minute more and up rose the ducks for the third time. They rose all but one. There was a flapping of wings on the surface for a moment, and then a duck went below. It seemed as though the bird had been caught by the feet.

I never saw the creature land with its capture, though Hyland thought he could make out a seal on the ice.

The drama was ended.

As we turned to go back Hyland shot a red-throated diver as it flew to the river from the sea. Sailor retrieved it—his lips tucked up and his whole face expressive of infinite disgust.

These were early days with Sailor; he soon changed all that.

And then we went to bed, for we had a long day's work before us.

. . . .

We rose at eight for breakfast, and then a strange thing happened.

We were sitting at breakfast when on the bank at the farthest point of this first reach of the river, distant barely 500 yards, there came into view, if our eyes did not deceive us, several persons moving down to the river. Presently they stopped, and one man sat on a stone and waited so.

It was an anxious moment.

It seemed to me that these natives, having sighted the tent, were suspicious and hesitating. They probably had their reindeer just behind the bank. What if, filled with alarm on seeing us approach, they should whip up and be off!

Were we to be cheated again?

I went for the whisky bottle, telling Hyland that the only chance would be to walk slowly along the cliff towards them and without our guns. For I thought that so, having time for a good look at us and judging us peaceable, they might wait our arrival in some confidence, being probably armed themselves. And the whisky, I knew, would seal our overtures.

We started.

We had not gone ten yards when something seemed to change. I raised my glass and took 'another long particular look.' And what do you think I saw?

Five bernacle geese.

Nothing more. There never had been anything more—except the mirage!

We set about preparations for departure; and you may like to hear what we carried on our backs. Here is the list:—

300 CARTRIDGE MAGAZINE, containing:

- 1 Biscuit tin of 43 lunch biscuits.
- 1 Biscuit tin of 39 digestive biscuits.
- 1 Tin of cocoa.
- 4 Pots of Liebig's extract.
- 6 Pots of Bovril.
- 4 Small tins of potted meat.
- Tin-opener.
- Kettle-lid.

SEXTANT CASE, containing:

- Dried apple chips.
- Dried vegetables.
- Raisins.
- 2 Desert spoons and forks.

CAMERA CASE, containing:

- Methylated spirits (pint bottle).
- Boiling machine.
- Podophyllin and quinine pills.
- Lint.
- Plaster.
- Essence of ginger.
- Cough tablets.
- Egg drills and blowers.
- Wadding.
- Pencils.
- Matches.

TIN BOTANICAL CASE, containing:

- Bacon, 1 lb.

We slung the kettle and two metal cups by the handles.

We took a change of socks, and so on—'for the evening.' I could not find anything better than a pair of canvas shoes, but Hyland had some Norwegian fur boots I had got for him in Vardö, and I am afraid I greatly envied him these.

Also I carried my telescope, flask and cup, pocket microscope and test-tubes, knives, watch, revolver, money, note-books, sponge-bag, soap, tooth-powder and brush.

Then we had of course our guns, and 125 cartridges each.

We weighed these things with the steel-yard, except the cartridges. Reckon each cartridge at 2 oz. and it comes to $15\frac{1}{2}$ lbs. a-piece; of food we had 18 lbs. a-piece; of extra clothing about 5 lbs., and our guns weighed between 6 and 7 lbs. each. Altogether we each carried a good big 50 lbs. weight, and rather more.

We intended to try and walk in india-rubber boots. But as I was very doubtful about the wisdom of this, we each carried at starting a pair of shooting-boots slung round our necks. These, however, we soon discarded.

Now this looks, as one reads it over, a funny list of things. But I think it is worth giving, for this reason, namely, that with the exception of the boots, some of the medicine, and many of the cartridges, there was literally not an item too much or too little. If I had to take the same journey again, there is not a single detail which could be added or be changed, excepting perhaps the clothes.

We were obliged to take all those cartridges, because we took them instead of food. For all we knew to the contrary, we might have to shoot our way back.

So I wrote a note for Powys in which I told him we were starting for Scharok. This I put in a tin and buried it on the cliff, six feet due north by compass off a cross. On this was written the words agreed upon between us.

Then we buried the axe in the bank, stowed the tent, bunged the barrel, and roped the tent down.

We were ready to start.

CHAPTER III

WE WALK ACROSS

June 24th (continued.)—We left exactly at seven in the evening; and we left in a roar of laughter. I began it, and Hyland took it up.

You remember the White Knight in *Alice in Wonderland*? Well, that was Hyland to a T.

Just try and picture, if you can, a man with a gun, dressed like this, and on a desert island. Start at the bottom and go up. India-rubber boots, cord breeches, a kettle tied round his waist with a bit of string. Two coats—a long-tailed shooting-coat, and over this a canvas Norfolk jacket covered with pockets of many kinds all bulged out to the utmost limit. Various things hung all about him, including a bundle of clothes, a pair of shooting-boots, a couple of grey skin boots, ornamented in red, with tassels; the whole topped with a flapped velvet stalking-cap.

I don't know how it reads, but it looked very comical, and my dress was much the same, without the skin boots.

First of all we made our way to the ford by the Russian cross—a distance of about two miles. But by the time we reached it, we had both come to the conclusion that a pair of heavy shooting-boots hanging round the neck was one of the worst devices. So we hid them. We laid them up by the grey goose's nest just above the ford.

The male peregrine was still hanging about the neighbourhood of the eyrie. Rising at our approach he mounted high into the air and flew round and round watching all our movements, and screaming wildly from time to time.

We were not to cross the river so easily this time. The water had risen.

Thinking it might be just a spate from the day's snow-melting, it seemed best to curl up for a sleep and wait for a fall in the flood.

It was not easy to find a suitable corner. The bank at this point dropped very suddenly to the water's edge where it was not wholly covered with snow, it was almost worse; mud in a semi-liquid state, slipping down in great treacle-like masses.

At last we found two places about forty yards apart, either of them capable of holding one man if he curled up close. The bank had become undermined and there had followed a land-slip, leaving a hollow ledge above. Hyland took one, I the other.

As I lay down there was a flutter under my head, and out of a hole slipped a little snow-bunting. So I turned my head and looked down, and there was her nest with six lovely little eggs. After that I just changed my position a bit, so that my head was clear of the nest. I was lying very still, watching a stone in the water which served as an index to its height: and I do not think more than five minutes had passed before she came, making no more of me than if I had been a stone, and—whir-r-r-p, —she had whipped close past my ear into the nest.

You will recollect that we had found these birds on Hornö with young ones able to fly, and here on Kolguev they had not yet hatched. Indeed these eggs were quite fresh.

A male snow-bunting in his breeding plumage is certainly a most beautiful little bird. Unfortunately we do not often see him like that in England, for the bird does not nest with us, or only sparingly, in the extreme north of Scotland. They are with us in winter in flocks, but then the beautiful black-and-white of the male has given place to more quiet colouring. Only when they rise from the stubbles where they feed, your eye is caught by a white flicker in the wings.

A ringed plover which had her four eggs a few yards off was not half so trusting.

It was very cold, but this silly little bird kept me often in most uncomfortable positions for long together, because she was so nervous. I should think she ran off and on to her eggs almost as many times in ten minutes. For she would settle down in quite a final way, but if I moved a finger off she went running about a yard or two away, and piping plaintively. Whether her eggs were spoilt or not I do not know. I think not. When I woke up about two hours later the first thing I saw was a little black-and-white head and a bright eye watching from the nest.

And then I looked at the stone in the river; the water had not sunk.

It must be a common experience with many who have tried camp life, or sleeping out of doors, that although they sleep as soundly they wake far more easily than at home in bed. They don't stretch or grunt, or pull themselves together, or wonder where they are, or protest in any way. Only they open their eyes. And with that, simultaneously, the thread of things is taken up again in all completeness. Quite alert, absolutely equipped, the sleeper is awake.

So the ringed plover was as still as the stone in the water for a minute or two after I woke; but then, when I was obliged to move, she left her four eggs to the mercy of the elements and ran piping off into the fog.

Such a fog it was—a sea-fog coming from the ice. And as it came it froze, and all it touched grew hard and white. My beard and moustache were solid, so that I had to break a way into my mouth.

Scrambling along the bank I reached Hyland's hollow. Poor fellow, I shall never forget him. He was lying there as white from the frost as the woollen sweater he was wearing. His face, ordinarily red and jolly, was blue with orange patches, and his hands the same. He could not speak, or only jerkily, because he was shaking all over. I thought he was in for ague, or had got it, and I fed him with quinine.

It was useless to reproach myself now for having taken him; I had to make the best of it. But I was very much concerned. I had never realised till then that people could feel the cold like that.

June 25th.—It was 4.15 A.M.: I could not let Hyland stop still, the more so as he seemed disinclined to move. So I got him up and moved him on. As he went he gradually shook off the shakiness, and by 7 A.M., when the fog cleared, seemed really all right, so we stopped for breakfast. Our allowance was four lunch biscuits each day; we now ate two apiece and spread on them some potted grouse. Also we made some cocoa, boiling the water with our methylated spirits.

We did not find a fordable place till noon. The river had been running most curiously in large horse-shoes, and always there was one side where the current ran hard and deep. But now it straightened out.

I felt very mean. In my long boots I was able to walk across. Hyland's boots were not long enough, and he had to take off his breeches again and wade.

At one o'clock we stopped for food and sleep. Poor Hyland, who was quite done up, dropped on the ground and was asleep in a moment. I was very anxious to husband the spirit as much as possible, so after throwing down my load I started out to see if I could not find something that would burn.

In the ordinary way you might as well look for—what shall I say?—a bishop at Rosherville, as for a piece of wood on Kolguev, excepting by the coast. By an extraordinary bit of good fortune, however, I came upon a small piece which had doubtless dropped off a sleigh. Carrying it back, I set to work to chip it with my knife. Every little chip (it was very hard, and would not split) I treasured as carefully as though it had been diamond—more carefully. At last, having made about two handfuls, and having piled up moss and green creeping birch, I really got a fire—a poor smouldering thing like weed-burning—but still a fire. Into this the kettle was stuffed. And after an incredible amount of blowing and choking and blowing again the kettle really sang. It wouldn't boil, but it did sing; a great concession under the circumstances.

I made some Bovril this time; it is safer than tea with lukewarm water.

This was a pretty good place. For there was a sort of little dry ditch which as you lay on your back just caught your shoulder-blades nicely and kept off the draughts.

So we slept there till 7.30 P.M. After this we rose another hill, and then came to a bad obstacle.

The hill formed a sort of amphitheatre. Circling round it, we dropped by steep slopes to the edge of a wide river. All the slope was covered with snow, which overhung the river. You could see as you looked sideways the river cutting underneath, black and swift under its white roof.

But it had to be faced.

I don't know how we managed to get down it, but we did; or at least I did, to begin with, while Hyland lay on the top with old Sailor and watched.

Choosing a place where a big stone just showed itself above the stream, I managed to drop on to it from the snow above. Fortunately, I am pretty light, but as I did so there was a sliding of the snow, and a huge fall followed me and went floating down the stream. But, at any rate, it had made a practicable breach.

Now, if you want to be comfortable, don't drop from a snow-roof on to a slippery stone in the middle of a swift-running stream whose depth you have not ascertained.

There was no turning back, for behind me was a snow-wall. I had to probe with my gun, and was glad to find I could touch the bottom. And then, very cautiously, I stepped down from the stone. The water just reached to within two inches or so of the top of my long wading boots.

Usually it is found an advantage to be a light man, occasionally it is distinctly not. When you are set to wade a very rapid stream, and the water is up to the top of your legs, the more you weigh the better. But though feeling as buoyant as Captain Boyton, and looking every moment to be floated off, I managed with the help of my gun to keep my legs, crossing the narrow channel, and emerging on a bit of land which just showed above the water.

And then I had quite a long exploration. For there was a second and deeper channel beyond. But finding my island tailed off into a shallow, I followed it carefully a long way down the middle of the river. At last, by little and little it led to the further bank, and I emerged safe and sound some quarter of a mile lower down.

But these river inquiries meant a double journey for me. Back I went till once more I reached my island. From the water it was possible to see better how the snow lay, and I found Hyland an easier place.

And now it was the old puzzle of the fox, the goose, and the bag of corn. How was everything to be got over? I could only settle it in one way. Balancing myself on my nasty wet little island, I pulled off the long boots and flung them across to Hyland.

It was an anxious moment.

The first travelled splendidly, spinning right up on to the top of the snow. But the second catching the edge of the snow-bank slithered down and was in the water before you could wink. You can imagine the desperate plungings and snatchings which rescued it. But it was rescued, emptied of water, and eventually figured on Hyland's other leg.

So we followed carefully along the track which I had taken and landed safely.

Now as this sort of crossing was often repeated, I need not trouble you with the manner of it again. Sometimes one could walk straight across the stream. But sometimes one had to follow a shallow a long way down, winning an inch or two at a time towards the crossing of the stream. Often these shallows were only narrow ridges with deep water on either side. Often you could only just manage it by feeling with your toes big stones on which you stepped; and, when they were slippery, down you went—a departure which tends to execrations. Sometimes, much to my disgust, just when I had accomplished successfully, inch by inch, a long and ticklish journey and was congratulating myself on a good crossing, lo and behold my little highway would suddenly end in a deep and impossible channel. But if all went well, then either Hyland would come across alone, or if it was exceptionally difficult, I would go back and give him a lead.

While we were sitting down to wring out my things Sailor amused us very much. For the old dog did not at all like this drop into the river off a steep snow-bank, and careered round for a long time in the hope of finding a better place. At last, finding himself perched like a chamois on the top of a steep snow cliff, he gave it up, ran back to our crossing, went bundling somehow into the stream, and came gallantly across.

On this river, the upper waters of the Isbushishna, we saw one pair of bernacle geese.

Most of my readers will be familiar with the name of this goose, and will connect it with the quaint old idea (which I have seen illustrated in a rare book) that it was hatched out of shells of the common barnacle which grew on trees hanging over the water. This bird, a winter visitor to our (chiefly west) coasts, breeds easily enough in a semi-domestic state; but where it nests in the wild condition has never been proved. Professor Collett, it is true, records the nesting for several years running of a pair on an island of the Lofotens, but this pair seem to have been as solitary as the instance.

So it would have been extremely interesting if we could only have found a nest. But we failed. This pair with our Gusina five exhausts our list of Kolguev bernacle.

June 26th.—We walked on through the night till 1.30 A.M. It had been a pleasant night. But at one in the morning a chill fog with wind came suddenly on.

After crossing the last river I had tried to keep to the high ground. But this was so much intersected by impassable gullies that we had perforce dropped down. The cold fog drove us up again on to a sandy plateau.

Here we constructed sleeping-nests. We worked like

rabbits till a good big hole was scooped out, and then built round it a fortification against the wind.

Now fine dry sand has its advantages, but also its drawbacks. The wind played with this sand through every chink in the rampart—built of lumps of saxifrage—and drove it home. By which I mean that sand—very fine—all but impalpable—went into your eyes, into your hair, up your nose, *through your trousers and socks.* I slept blindfold, with a handkerchief tied round my head. Hyland, who did not take this precaution, suffered much from his eyes for a day or two afterwards. I suppose it must in reality have found its way into our lungs. Anyway, our throats, on our awaking, were full of sand.

We found nothing here with which to make a fire, so we had to boil water with the spirit-lamp, and making some cocoa, ate a lunch biscuit each and a raw rasher of bacon. We were obliged to be a bit careful with the biscuits on Sailor's account. As it was, the poor old dog got very short commons.

The country through which we now passed when we moved on at 8 A.M. consisted of a wide stretch of high-lying mosses and swamps containing many enormous beds of peat. As I shall have occasion later on to refer more particularly to these peat-beds, I will here only remark that many of these were of most curious formation. Some perfectly circular, many absolutely rectangular in shape, they rose suddenly from the bog or water with perpendicular sides to the height of four or five feet.

At 11 A.M. we had lunch on a high, barren hill, and, crossing a river which ran at its foot, headed for some mountains which loomed high in the distance.

These mountains looked as if they might be at least 1000 feet high. But long ere this we had learned to mistrust the evidence of our senses in this strange land.

A story such as that of the five bernacle geese (told earlier in this account) might well meet with a smile of incredulity from those who have not themselves been in Northern regions. But the following quotations will show that ours was but part of the common experience of Arctic voyagers.

One morning, when we were in Mossel Bay, we were roused by the watch at an unusually early hour, who announced that there were twenty or thirty deer by the shore washing themselves in the sea. Field-glasses and telescopes were brought to bear upon the dark moving objects, and two boats were lowered immediately to take ashore the three eager sportsmen and the captain, all anxious to secure some of these, the first reindeer we had met with. Ascending to the crow's nest soon after they had left the ship, I saw through the ship's glass what was taking place, and anticipated the results of the drive. It was evidently nothing but a wild-goose chase; so I made my way down and announced that there would be no venison for breakfast that morning. After they had rowed some miles, the hungry sportsmen returned empty-handed; but they had *seen* the geese.[1]

[1] Rev. A. E. Eaton, *Zoologist*, Nov. 1873.

And again :—

It happened once during foggy weather on the ice at Wahlenberg Bay that the bear that was expected, and had been clearly seen by all of us, instead of approaching with his usual supple zigzag movements, and with his ordinary attempts to nose himself to a sure insight into the fitness of the foreigners for food, just as the marksman took aim, spread out gigantic wings and flew away in the form of a small ivory gull. Another time during the same sledge journey we heard from the tent in which we rested, the cook, who was employed outside, cry out, 'A bear! a great bear! No! a reindeer, a very little reindeer!' The same instant a well-directed shot was fired, and the bear-reindeer was found to be a very small fox, which thus paid with its life for the honour of having for some moments played the part of a big animal. From these accounts it may be seen how difficult navigation among drift-ice must be in unknown waters.'[1]

Well, we were constantly taken in in the same way.

To-day, for instance, we were just coming up, as we thought, to a large lake when, turning to Hyland, who was following some little way behind, I said, 'Look there. That is the first swan we have seen since we left the Kriva.'

'So it is, sir,' answers Hyland. 'We must try and shoot that.'

I took the 'number 6' out of my gun and slipped in a brass cartridge with S.S.G.

Presently—'That's not a swan, but an old cock willow-grouse with his head up,' I remarked. Hyland assented. (Polonius and the cloud.)

[1] Nordenskiöld, *Voyage of the Vega*, vol. i. p. 348.

But again I had my doubts. Indeed, the bird was changing very strangely. And then I saw plainly enough that it was not a bird at all, but an Arctic hare sitting up as hares will sit. I was much excited about this. It was the first hare I had seen.

I crept up. The hare sat on,—still a hare, but dwindling. To cut a long story short—when I came within shooting distance it was to find neither lake nor swan, neither willow-grouse nor hare, but a little cock snow-bunting perched upon a mound against a drift of snow.

Further on on the top of a wind-swept shoulder of the hill I took four eggs from a grey plover's nest. The nest was a deep circular depression containing, with the exception of a little lichen, nothing but the eggs. The hen bird I shot, but the male was impossible to secure, he was so wild and wary. After waiting about a long time for him I had to give it up.

As this bird flew off it was met by one of a pair of Arctic skuas who were hunting the ground for eggs. The skua made a cut at it. But the grey plover, rising high in the air came down on the unlucky skua like a bolt, and followed him up, wheeling and buffeting in fine style.

We reached our mountain to find it was about 150 feet high.

It consisted of three conical hills, rising very suddenly, and grouped in the shape of a triangle. The base of this was open, but the hills were tied together on two sides by a high sandy causeway. At the bottom ran a stream.

The lower part of the formation was clay, now in a semi-liquid state ; but the tops were of sand.

I had not intended to stay here. Really, I had mounted the hill in order to try and get a view of the surrounding country. But a dense fog came on as usual, and so we stopped.

We found it very difficult to choose a place for a bed ; for all the hollows were filled with snow, and the north wind curled keenly in every direction.

At last we contrived, by taking advantage of a slight depression, a tolerably protected retreat with moss and turf which we cut with our knives.

June 27th.—I wished much that I had a thermometer to register the cold of last night. It must have been very considerable ; for the little stream which when we came was racing like a cataract, was, but for a tiny trickle, stopped. The fog, too, at midnight was very dense, and drove along the sides of the hill like whirling desert dust.

I must confess I envied Hyland his warm skin-boots. Indiarubber is the very worst material for a cold country. I kicked my feet about to no purpose, so at last I pulled off my long boots and tried my canvas shoes. They were better.

At half-past twelve a sound woke me up. Hyland

had another ague attack. Fast asleep, he was shaking all over, his teeth chattering like castanets. I knew he was dead-tired, so concluded that he had better sleep a little longer. I threw over him a thin coat I had—the only thing I could spare—packed him all round with moss, and climbed the hill for a walk. I collected a few flowers and lichens, and returned, for there was no chance of a view.

At 2.30 A.M., on waking up, I found not only that my moustache and beard were solid, so that I could not open my mouth, but that my eyes were also sealed by ice. After a somewhat painful operation I managed to get them open. The fog had cleared a little, and again I went for a walk.

Traversing the sandy causeway, I made for a high point, but was brought up sharp by a ravine which seemed impassable.

But as the fog lifted here and there, I could make out that our hill was just an isolated rise, and that in front of us lay a vast plain which we should have to cross.

Poor old Sailor, who had hitherto been tolerably independent and regardless of cold (for he is the hardest dog, I think, that I have ever seen), was fairly beaten this time. He crept close up to us for warmth. Hyland, too, made a grand discovery. It was this, viz., that your extra socks are a very fair substitute for gloves. It seems extraordinary that one had not thought of this before; for our hands had often been exceedingly cold.

I may mention here another discovery worth remembering. When you can find no shelter of any kind, and are not well clothed, the discomforts of sleeping in the open air are much mitigated if you sleep with your head to the wind. For it is the back which suffers most.

It was just four o'clock that morning when, after having eaten a fig and a lunch biscuit, we were once more on the move. The fog had cleared, and the sun shone most brilliantly.

I have somehow forgotten to tell you about the flowers; but as I mean to give them a separate chapter later on, perhaps I need say but little here. The most beautiful of them all, I think, was the Arctic Forget-me-not. Of an intense blue, it flourished but on the sandy places. Commonly it grew on the pure sand, unsupported by any other green thing. Here it looked strangely unnatural, I used to think, like the flower in a lady's bonnet.

At 7 A.M. we reached a little half-frozen stream. I got out my sponge-bag and made a toilette with some difficulty. Then we had breakfast, finishing our first tin of potted grouse. And so we went to sleep.

Just after we had started again at 10.30, I all but trod on a long-tailed duck. Bang off her nest she went, scattering her eggs in her flurry. She had six eggs, and they were slightly 'set.' The nest was remarkably neat, and, for a duck's nest, very deep. It was all of down with a little dead grass and dead birch-leaves.

We walked very slowly. To understand how this

was you must put fifty pounds weight odd on your back and with it walk a mile or two, sometimes on Brighton beach, and sometimes in a bog which takes you almost up to your knees.

When we stopped for luncheon at 3.30 P.M., we each ate a lunch biscuit spread over with Liebig's Extract, and a raw rasher of bacon, and drank snow-water. We couldn't make a fire, and felt obliged to save the methylated spirits.

Lying on my back, in a half-sleepy state, after luncheon, I was much interested in watching the ways of a pair of snow-buntings.

They were building a nest out of sight under the overhanging snow which formed the bank of a tiny stream. The hen worked exceedingly hard, exerting all her little energies to pull fibres from the ground. The cock was a fraud.

You know how some persons make a great show of passing things to you at luncheon-time. Actually they do nothing—they let it go. But each time you reach for the salt, they wave a hand after it as though a magic wand. The cock snow-bunting had learned this trick well. Very assiduous, very fussy, he accompanied his wife up and down. He waited upon her while she hunted up fibres; he flew back with her and watched while she worked at the nest—a perfect example of a despot lord. I do not believe he gathered a solitary thing himself; I never saw him do it.

A shore-lark's nest, with six eggs, which I took on this evening, was entirely composed of dead grass and willow down. This bird, which the Samoyeds call 'sĭzīŏ' from its note, is another of our Arctic visitors. It nests in Greenland and Novaya Zemblya, and all about the Siberian and Arctic Russian coasts. It has an exceedingly pretty song, about which I shall have more to say later on. Its plumage is very pretty, for its throat is yellow, and its breast broadly banded with black.

And then at 8.30 P.M. we found the nest of a little stint with four eggs; far away from any water, for all that we could see. The bird twittered and ran about close to us, feigning lameness. The nest was a deep cup in the ground, half filled with dried birch-leaves.

Now, exactly at ten o'clock in the evening, we came upon a little lake. A good dry bank sloped down to its edge, and on this we lay very comfortably with our backs to the wind.

We were pretty hungry. After eating a bit of raw bacon and a biscuit we made some cocoa. I had fenced the spirit-lamp with lichen and moss as a guard against the wind. Presently the moss itself caught fire; and this idea flashed upon me, Why not make a moss fire and try to cook with it?

Hyland did not believe this possible But I set to work and scrabbled up all the moss round about, earth and all—it was very dry—and presently had a good big smouldering fire.

Now we had a cock grouse; and the next and obvious reflection was, Why not cook a grouse?

So Hyland cut off the wings, pulled out the tail, cleaned the bird, and then we stuffed him, feathers and all, into the middle of the fire, and collecting more and more moss piled it on the top.

From time to time we took a peep just to see how the bird was getting on. But always he was just the same feathered fowl as we had put him in. The feathers were a trifle singed, and that was all. At the end of some half-an-hour it looked just as hopeless. This we could not understand, because the fire really was quite hot inside. So then we began to probe with knives. They seemed to go in very easily.

There was no doubt about it—the bird was getting soft!

We gave it another ten minutes or so, and then pulled it out of the fire, determined to have grouse for supper, done or underdone. It came out a perfectly cooked grouse. The feathers, which were barely singed, stripped clean off with the skin; and there he was, just like a gipsy's hedgehog.

I can only tell you this, that no grouse ever turned out of the kitchen by your best fancy cook was a patch on this bird. And that it should be so stands to reason. For trussed and roasted in the ordinary way all the tea, so to say, runs out of the tea-leaves. So that I am sure ours is the way in which to cook all feathered game.

This night we did not sleep. I have never been out on a lovelier night. There was not a trace of fog. Clear as it had been all day, the wind died down at morning. And over the little lake, half covered with thick broken floes, you could look, and far away across the tundra, for there was no mirage now. Grey and purple the peat waves of the tundra rolled on and on, and the sun as it rose higher (it had never set, remember) touched the little distant tarns till they twinkled like the dewdrops on the lawn at home, or drew off the top of the marshes small soft clouds of white.

We lay under the shadow of the bank while the further side of the lake was lit with sunshine. It was all worth looking at. I felt it was not a bad thing to be houseless on Kolguev just then.

And soon after midnight, just as the lake itself began to steam, we had a visitor. Down from the top of the bank on the further side jauntily stepped a little fox, and sat looking down at the floe below him. He was patchily coloured blue and white; 'christovatik,' as the Russians call it, because the dark mark on back and shoulders shows the form of a cross when the animal is skinned.

June 28th.—We moved on at 5 A.M.

You will perhaps recollect that all this time we have been steering for Stanavoi Scharok. But last night, after thinking over the whole position, I had changed my plans.

In the first place, we had already crossed twice a river which must be of considerable size as it made away to the east and south. What river was this, and where did it debouch? Neither of these questions could we answer then, though later on we knew it for the Pesanka.

The Baroshika indeed was, according to the chart, a biggish river. What if we got down into those plains which we saw and found this river wide, deep, and with many horse-shoe windings? And then Hyland, and his sufferings from cold, was always in my mind. Drift-wood, drift-wood, at any price it seemed, was the first aim before us.

And so this day I went a point more easterly, aiming for where a beacon on the sands had been indicated on our chart.

My little pocket-compass was rather a worry. Not only was I obliged to put away my gun each time I consulted it, which was pretty often, but unfortunately the needle had a habit of sticking; and this, unless you were very careful, would escape notice.

But I found the sun exceedingly useful as a means of checking our course, by what I may call the '*shadow* method'—which expression, I take it, explains itself.

I have found it almost impossible to make those friends here at home who have not been to these parts realise that there is a real and very beautiful *summer* in the Arctic Regions. The cold fogs and frosts we had experienced were, of course, very trying, but we should have been all right if we had had sleeping bags or a tent, or even good warm great-coats. One cannot expect to be very comfortable anywhere where it freezes when lying out at night. It would be just the same in England for that matter. Try it. Some November day when the last beat is finished in the home coverts, and the other guns go back to tea, try it. Take your gun, stroll off, and lie down for the night in a furrow in the middle of a ploughed field. You will not find your eyes frozen up, but otherwise, by the time the morning breaks, you will have gained a very fair idea of a night on Kolguev.

But this day we suffered from *heat*.

It came on gradually about ten o'clock with a sort of muggy, Torquay feeling.

We began to feel our loads as we had not felt them before. From something Hyland said I fancied he thought his load was heavier than my own. So we changed loads. If we had been able to contrive a proper means of carrying these we could have managed that weight splendidly. Another time I should be strongly inclined to use one of those fish-creels which the women carry in the North Scotch towns. This would take everything, and would put the weight in the right place. As it was, we were all straps and cords, which cut on our shoulders and dragged on the neck and spine. My long indiarubber wading boots, though very important for the rivers, really handicapped me considerably, especially in the bogs.

As the day wore on the heat became intense. It blistered our faces, and our hands swelled up like puddings.

We could now only advance by stages. The plan was this. I, trudging along in front, would doggedly count about five hundred paces, and then sit down wherever there was a mound which would take the weight of the cases off our backs. That was a great relief. After five minutes by the watch up we would get again and plod along.

Now one remarkable fact in respect of the intense heat was this, that neither of us turned a hair. How was that? I do not know. But so it was.

Sleep, too, was out of the question, because a new thing had happened. Musquitoes had appeared.

But there was no help for it. We were heading for some hills that would be about eight or ten miles off; and in the meantime all the land was weary bog with peat-mounds here and there. Poor old Sailor suffered badly from the musquitoes, which bit him about the muzzle, so that he had constantly to put his nose into a bit of moss and rub it with his paws.

At last we reached the hills, and after a long and weary climb found ourselves on a high plateau. Here there was a bit of a breeze, and the mosquitoes were much less troublesome.

Hyland was asleep before I could feed him, and after I had smoked a pipe and had eaten some raisins I followed his example.

We slept for an hour or two, and then ate some cold Liebig, potted pheasant and biscuit, and trudged off.

These little tins of potted meat were all labelled differently, but all tasted the same.

We went east for ten miles, as near as I could estimate it—putting our average at two miles an hour—and at 9.30 P.M. reached some sandhills with ridges of stones, which were evidently old beaches.

To-day we had seen a snowy owl—the third. Like the others, it was a dark bird. It was sitting on a hillock some distance off.

The last two miles of our walk was through a most lovely cloudberry garden. I can call it nothing else. The bog was filled with mounds and ridges of peat, all white with this beautiful flower. It was like a splendid strawberry bed, but with more flower and less leaf.

We came upon a pair of Arctic skuas who were evidently nesting, and tried hard, but unsuccessfully, to find the nest.

Here was the highest growth of vegetation we had yet seen, for the woolly willow was at least a foot high. In the marsh I also found a Lapland bunting's nest containing seven eggs. The nest was composed of grasses lined with feathers.

I am afraid I have in some instances forgotten my

undertaking to say something about the birds as British. The Arctic skua, a dark form of which is known as Richardson's Skua, belongs to a group of sea-birds of great interest. Four of these occur on our coasts, but only two nest with us, and on our extreme Northern Islands. The great skua, the 'bonxie' of the Shetlanders, is one, and the other is the bird above. They are the 'robber-gulls;' indeed, when one has grown familiar with their ways in such places as Kolguev, one comes to think of them as raptorial gulls. Not only do they rob at sea the poor little other gulls, but quarter the tundra for eggs and young birds as steadily as a brace of setters or a pair of marsh-harriers.

A light wind was blowing, and it was not very cold nor very foggy. The snow, in melting off the top of the ridge, had worn a little water-course which was now dry, and here, by pulling up lumps of saxifrage, sand and moss, we were able to construct very comfortable nests.

The ridges too were covered with creeping birch, most of it dead and brittle. After working for half an hour or so, we had collected quite a big heap of this and made a capital fire.

Here, in a pool, I shot a long-tailed duck, and Hyland a willow-grouse off the grasses, for to-morrow's breakfast.

After spending some time in writing up my diary, and in copying out my route observations, I set to work to collect every stick I could lay hands on for the morrow's fire. I worked a long time at this and wandered far. I moved a grey goose [a bean] off her one egg, and then I grew contemplative. It was a lovely night, and I began conjecturing—'Now, apart from the position of the sun, what is there to show you it is not day? If at this moment you were suddenly dropped down here, and did not think of looking at the sun, would it strike you, in any way, as a curious day? What, in short, is the distinctive characteristic of this Arctic night?'

I think this was the beginning of it; but, at any rate, I got into a brown study and went wandering on until I was lost. I really was. And I stood and laughed at myself for being such an idiot. But all the hills were just alike, and our little crevice was so low down among them, that it was only by making a cast and hitting off our track that I found it.

But at two o'clock in the morning I curled up. The last thing I saw was a pair of old glaucous gulls come flying in from the east.

So I went to sleep happy, knowing that the sea was not far away.

CHAPTER IV

REACHING THE GOAL

June 29th.—But we were not to have it all our own way. A biting north wind set up, and at 6.30 A.M. we rose, made a fire, boiled water for cocoa, finished the potted pheasant, ate some raisins, and then went down to the marsh to try again for the skua's nest.

The way these birds 'carried on' passes all description. They tried to lead us away from the nest by every conceivable device. They pretended that their eggs were in two or three places other than where they really were. One very striking phase of the performance was the following: A bird would drop in the water as if shot. Then it would flap helplessly for a bit, and if this did not move you, it would raise itself on its tail, beating forwards slowly with its wings and mewing like a cat. 'Mewing' exactly describes the sound.

Sailor was very funny about it. They had a most tempting way of struggling painfully along on the ground about two yards in front of his nose, so that he was quite sure that he could catch one if only allowed to try. But, rated soundly for breaking once, he did not dare to try it again. Only he went crouching along, treading as softly as possible, in a quiver of suppressed excitement, like a cat preparing to spring.

There were two eggs in the nest when we found it at last.

The Lapland bunting is a little yellow-billed bird, with a black throat and a white boa, which never nests with us, and has only been taken some half a hundred times in England. But it is a very common bird indeed on the tundra, and, with its plaintive little piping note, was our constant companion.

But luncheon-time came, and again we cooked a willow-grouse in a moss fire, and again with successful result.

While I was sitting down and sketching the scene a couple of geese flew over. I snatched up my gun and dropped one of them. It was a male bean-goose, and I am afraid the parent of the egg I found early in the morning.

We had never before found such a pleasant camping ground as this.

From the little crevice in which we lay ran two dry gullies, down which we could walk unseen to the pools in which they ended.

In one of these I shot a male long-tailed duck. He had a very good tail, but was otherwise disappointing, for his head was dark. So we ate him. Indeed, since we left the sea, we have not seen a single male of this species in full plumage.

I have grown as tired of writing, as you, I am afraid,

must be tired of reading about this walk. So we will hurry on.

At noon, then, we got under way. Crossing a small plain, we rose some exceedingly difficult hills. It was a constant struggle with snow defiles, with sudden drops and weary climbings, or with sinkings deep into mud puddings as a little bit of change.

Old Sailor, who left the ship as fat as an alderman, for he was a sad galley-loafer, had by now worked himself into hard condition, though he was as thin as a rake. But to-day, what with the long-tailed ducks, willow-grouse, and goose of his mid-day meal, he had a load which he found it very hard to carry over some places. More than once he lay down nearly defeated.

I ought to have said that on this morning, when washing my hands in a stream, I saw a small water-beetle, shaped like our *dyticus*, swimming about; but though I tried to catch it, it eluded me successfully every time. After all, we could not have preserved it, I fear; for the boat had gone off with all the bottles, spirits, and appliances of that kind.

Now it chanced that at half-past nine in the evening we were struggling through such a country as this, when, on the top of a distant hill, a little peaked lump came into view. Instantly my glass was at my eye. The sling telescope which I carried had been used many times a day. For at first we were always supposing that we saw persons and reindeer. But the mirage was such that very soon we had learned to distrust these appearances, which the glass too often showed to be nothing better than a heap of stones or a ridge of peat.

And just as once I should have only looked to confirm my certainty that here were human beings, so now I only did so to make sure of my belief that here were not.

You can therefore imagine a little of my astonishment when the glass showed me this time, as it really seemed, a native choom and many reindeer well defined against the sky.

Now, when you reflect upon men in our case, it would seem an intelligible impulse to have danced about or cheered, to celebrate this great moment. But then, I thought, what if, after all, this should be but a creation of the mirage as others had been before! It would be but a cruelty to have raised my companion's hopes only to dash them again.

These considerations made me careful, so that I only said, 'We will stop here and wait a moment.'

And then, having looked again with a steadier glass, I saw with such clearness the choom, and the reindeer as they moved along the sky-line, that I could no longer doubt. But when Hyland at last asked me what I was looking at so intently, I said quietly and very casually:

'I fancy I see a choom and many reindeer.'

For now that I was certain it seemed pleasant to play with the occasion, nor would I startle Hyland into excess of joy. But, owing to the tone of my voice, this news affected him with no greater sense of credibility than had I said, 'I see the Lord Mayor and the Bank of England.'

With this I gave the glass to Hyland. And no words of mine could convey the intensity of his delight on really seeing with his own eyes that this wonder was no illusion. For just as we have all known many a fisher to mistake a weed or log which he has struck for a pike, but have never known the contrary happen, so, for all the deceptions of the mirage, when the real thing came there could be no mistake. There was nothing for it now but to march right on till we came up with the choom.

I had always in my mind that old account of the Samoyeds on Yalmal, who fled away from their Norwegian visitors; which made me suppose that in such an isolated spot as Kolguev, where visits of the foreigner were all but unknown, we might find with them the same experience, which would be a grievous disappointment.

However, I need not have felt this concern, as will presently appear. So we kept plodding on, over one of the most difficult bits of country with which we had yet met. Up the hills and across the gullies we struggled, until at last I said we should make some tea.

For I experienced that which many others in similar cases have known before. As long as there was imperative need for pushing on I never seemed to be really played out. Always, as horsemen say, I had a leg to spare. But now that there was no instant occasion for haste, now that we had really found that which we had been seeking, it was just reversed. Hyland was grown quite fresh and filled with new energy in this discovery, while I had found that I was really tired, and that the constant carrying of this heavy weight had in some way told on my spine.

So we stopped half-way up a hill, and made some tea, using for that our last drop of methylated spirits.

At one o'clock in the morning we came among the deer. To my surprise we found that a large proportion of them were white. Meantime a very heavy fog had settled down upon the hills. We could not see with any certainty more than twenty yards in front of us. I had been sitting down to rest, and on rising again to move away my companion remarked that I was going in the wrong direction, and that the choom was just the other way. And indeed I had done a foolish thing. For, since sighting the choom at first, I had put away the compass for good and all, knowing that to walk straight up to it from that point was only a question of keeping a clear head. But no sooner had we entered the hills than we lost all sight of the choom and the reindeer, and had not since seen the choom at all.

And as I had been lying on the ground with closed eyes, I supposed that I had become confused, and therefore followed Hyland's lead. But it did not answer.

We came upon no choom. The fog grew denser and denser, and at last there was nothing for it but to make a cast round. This showed us nothing, and it really seemed as though we had missed it altogether.

And now I am going to relate a fact which I cannot explain excepting by a parallel taken from the physics. Just as by the first law of motion a body will continue in motion in a straight line unless acted upon by an intercepting force, so it seems it must be with the brain. I cannot help believing that to this must be greatly due the notorious faculty which savages have for finding their way from point to point. I do not mean that such agencies as sun and wind have not a determining influence, for of course they have; but that, apart from this active and observant intelligence, there is a consciousness of direction which is obeyed subjectively. It is instinctive, as we say. I remember once asking a Cree with whom I was hunting how he found his way home so wonderfully well, and was much struck by the man's answer. Putting his hand to his side he said, 'Indian's *heart*.' I thought it was such a capital answer. By it I suppose he meant that the Indian's affection would naturally be a sufficient guide to the home to which it pointed.

The heart with him was the seat of the 'homing instinct.'

And so it doubtless is. And this function of the mind [whether 'the heart' (*i.e.* the affection) or any other motive of desire] is served continually by the intelligence so long as it remains unbroken, uninfluenced by some other force.

We say, then, that we 'know the direction,' and this consciousness is such that if its object is retained and pursued day after day the mind becomes so trained in a certain line of concentration that after a check it can take up the thread again quite independently of the eye.

I remember, again, how one day when it was very hot I threw a jacket over the bough of a tree where my Cree Indian said we should pick it up on our return. We hunted all that day and the next, lay out that night, and the following evening as we were returning home I suddenly remembered my jacket. A jacket hanging on a bough in the middle of a pathless forest is no very easy thing to find. But the Cree, shutting his eyes, remained so for a minute or two, and then, turning half round, walked straight away. I followed, and at the end of an hour or so we had come straight up to the jacket.

This has been a long digression. But now to the point.

I remembered the Cree. I shut my eyes and thought. After a moment I could, as it were, *see* the choom, and so clearly that I knew I could walk up to it.

I opened my eyes. All was fog—dense fog. But pointing, I said to my companion, 'There is the choom—straight over there,' which was almost in the opposite direction from where we had supposed it to be.

We shouldered our things and marched on, and sure enough it was not long till we saw before us the dim outline of the choom looming through the fog.

Sailor, too, knew we were approaching something strange. Ever since we had come upon the reindeer ground he had been very uncomfortable. Unlike himself, usually so cheery, he had dropped his tail and advanced in a doubtful, hesitating manner. And now he was so staggered that he kept stopping, and we had much ado to get him on. No doubt it was because of the smell of reindeer and of strange dogs. Sailor was never afraid of these dogs, but braver than them all, as will appear in its place.

CHAPTER V

FRIENDS

At half-past three in the morning we reached the choom, a circular peaked dwelling covered with birch bark. No one was about, but a deafening chorus arose from many dogs who were tied up round about it.

'How do you do?' I said in my bad Russian as I raised the flap of the choom door. 'How do you do?' came back in Russian worse than mine from underneath a heap of skins. The first impression one had was that the whole floor was a mass of skins, with a pot hanging in the middle; nothing remotely like a human being could I see.

And then the floor of the choom began to bubble and rise, so to say, and out of the bubbling came a little man and a little woman all clad in skins. On their heads skin hoods, on their bodies skin caftans, on their legs skin boots. They were dressed much alike, but one had the smoother face and was decorated with bits of red. So I knew it was a woman. She was plain and wrinkled, but had a not unkindly twinkle in her eyes. They waited as patiently as a pair of public officials for us to explain ourselves, only they gazed from one to the other with bewildered looks. In my bad Russian I did my best. It was a comical conversation.

'You Samoyed?'

'Yes yes, Samoyedi—Samodine, yes yes. Nyanitz, in Samoyed *Nyanitz*.' So we had already learnt the native name of the people.

'We Englishmen'—(we might as well have said 'We Angels,' for all they understood). 'Englishmen from the Gusina.'

'Ah, Gusina,' said the woman quickly. 'Yes, yes, Gusina, Gusina, yes, yes,' the man went on. These people have caught and amplified the Russian repetitive.

'We walk on foot from Gusina.' He didn't understand, so I stamped round. Then he grasped my meaning.

'No, no, not possible. Much hardt. No, no.'

'Yes, yes, very hard, but quite true.'

(But he didn't mean that. 'Hardt' is Samoyed for *snow*.)

Then I gave them both a little whisky, for I had a flask that held a wineglass and a half. This put us all on better terms.

I told him he must take us back to get our things; that my choom was there. At this they laughed, but said 'Impossible. Reindeer ill.'

So I showed the Governor's letter. The spread-eagle they recognised, nothing else.

There was something about these small people, absolute masters of all we saw, that gave me an uncomfortable sense of being an intruder. We in our English clothes, with our English guns and English spaniel, seemed as strangely out of place as a tramp in a West End drawing-room. And frankly, I think I should have scarcely been surprised if the man had said in superior English—'Well, my good men, I'm sorry for you, but really I've nothing to give you. I can't have tramps coming here.'

However, while this was going on we had entered the choom. Various sleepers moving and grouping themselves on one side left the other free for us.

They lit a fire and the women cooked us a goose, cutting it up and throwing it into a great pot. It was excellent.

At first the smoke of the fire was very disturbing. It got into one's nose and eyes—but we soon found that by keeping low down, as did our entertainers, the smoke was little felt. They were much amused, laughing heartily when they found that we could not sit comfortably cross-legged on the floor as they did; and presently gave us tubs to sit on.

UANO

The choom contained:—

Uano, a Samoyed, about fifty-three years old.
Ustynia, his wife.

Mekolka, his son, unmarried.
Ustynia, } daughters.
Anka, } daughters.
Tierrtso or Zornka, } daughters.
Katrina, his daughter-in-law.
Niab-kutni, her little girl.
Wanka, her little boy.
A baby.

I have given you all the *dramatis personæ*, just as they were introduced, or rather explained to us by Uano, except the three last who did not come out of the skins. This was pretty well for a circular room about fifteen feet across.

While the goose was cooking we had time to consider our entertainers more carefully. There was a touch—but not more—of Mongol in all their faces, more marked in the women than in the men. They had high cheek-bones and a tendency to slit-like eyes, and their eye-brows were beautifully arched.

Uano had a short grizzled moustache and beard, and a world of shrewdness in his eyes, which were redeemed from cunning by a kindly twinkle. His wife Ustynia, whose old face was as yellow as his, was not ugly. Just a homely, kind-looking old woman who would have made a respectable cottager on any English estate.

Katrina was almost pretty; indeed when first she married she must have been a pretty girl. Her eyes were dark, she had a pretty colour in her cheeks, and her manner betrayed a quiet self-possession very remarkable among such a people.

Her sister-in-law Ustynia was really, if you accept the type, a pretty girl. Her broad forehead was banded by a circle of brass ornaments and beads, which keeping back her black hair wound about it as it fell in two long plaits, and finally about her waist, divided into a triple row of heavy beads joining the two ends together. This is the common fashion with unmarried women. And I think if English girls knew how pretty and becoming this forehead band is they would wear something like it too. Ustynia's eyes were bright, and a pleasant smile played about her lips. When she laughed—and these people are always laughing—she betrayed the most perfectly beautiful teeth it is possible to imagine. Indeed all these people, even old Uano, had most wonderful teeth—white, regular, and perfectly shaped. On her fingers Ustynia wore heavy rings of white and yellow metal, and her hands, like those of all Samoyeds, were faultless in shape and extraordinarily supple. If you add to this a dress reaching to the knees, formed of young reindeer skin, worked in many stripes of white and brown, the skirt—(the dress is all in one piece and fastens in front)—banded with scarlet cloth and dog-skin fur, and foot and leg coverings of soft, patterned skin reaching above the knee—there you have Ustynia, the belle of Kolguev.

Anka was a replica of her sister on a smaller scale. She would be about fourteen years old.

USTYNIA

Face page 130

They served our stewed goose in wooden bowls, and laughed much at our efforts to sit cross-legged in their fashion. I made them some cocoa, which they said was 'very, very good. Yes yes.' Then old Uano begged my pipe. There was no pipe on the island, he said, and so I believed, for he let it go out every other moment. Everything we had, our matches, our spoons and forks, our boots and guns, were a wonder and delight to them. All were critically examined, and they wanted to have them all.

Well, we laughed and talked till 3.30 A.M., then lay down on our reindeer skins for sleep. Hyland was asleep in a moment. But I lay for a long time smoking my pipe and taking in all the details of this curious place in which we lay. Then I too curled up, determined to sleep as long as ever I could. For I felt we deserved it. The first part of our endeavour was at any rate achieved.

We had found the Samoyeds.[1]

[1] (1) Just as the Ostiaks call themselves 'Habi,' so the Samoyeds call themselves 'Nyanitz.' Also 'Nyanitz' is a Samoyed *man*, while 'Nya' is a Samoyed *woman*. The Samoyed name for a Russian woman is 'Habynia.' They never say 'Baba.' But very curiously the *mainland* Samoyeds, when pointing out one of their own women to me, said 'Barena,' *i.e.* 'Lady,' as opposed to 'woman.'

(2) I have just been reading in my journal a 'Review of our walk.' It is too long by far to be quoted at length. But the following is curious, I fancy :—'The result of these various influences (cold, heat, frost, etc.) is a remarkable swelling of the hands and ears. Our hands are huge and puffy, the skin quite tight, and very, very dry and brittle. Almost every day I knock off a small bit of skin by some slight accidental touch, and it does not heal or get worse or "angry"; only it slowly widens to a cut or crack. I have on my forefinger now a slit which began as a little chipped-off bit of skin (on the ship), and has gradually widened and deepened—it is just on the first joint—till it forms a clean deep cut across the finger and nearly to the bone. This is very interesting and curious. I think it is the effect of the extraordinary dry wind. The fact of our not perspiring under a heavy load in the baking sun is in itself worth noticing.'

'The exasperating feature in this week was that it was *impossible* to do much collecting. I made notes, picked what flowers and took what insects I could; and when we halted searched the banks for Saurian and Mammoth remains, though I found none. But any proper work was impossible. We could not carry an ounce more, not even eggs; though I did manage to get the grey plovers' and the little stints' safely through. I was struck by the dearth of mammals, the dearth of raptorial birds, the absence of swans, the limited list of passerine birds. Of these two last we have only added the white wagtail to those we saw before.

WINTER HOOD
OF SAMOYED WOMAN

PART III

Shows how we re-crossed the island with sleighs and many reindeer—Of the wonders these animals performed—Of the manner of life of the Samoyeds—The choom, the sleeping, etc.—How we met On Tipa, the rich man—Of the wolf-like dogs, of knives and boats—Of the crossing of a river, with other matters domestic and not so—And finally, of how we travelled to Scharok harbour to wait for the *Saxon*, which yet could not come for the ice.

CHAPTER I

RE-CROSSING

June 30th.—We were now in—though we did not know it—for more than a three months' stay with the Samoyeds.

When I actually came to live among the Samoyeds and to see all their manner of living, I found all so different in many ways from the books, that I think a good deal that has been written must have been taken at second-hand.

It is possible of course that I have—nay, I must have made some mistakes as to meanings and significances—they are always hard to be sure about—but facts are facts. And you will have them here at all events, in an accurate record of our life among these interesting people as we knew them.

You know what winter hop-poles look like when stacked ready for putting out. Well, cover them round with birch bark and you have in appearance the outside of a choom.

The word 'choom,' or *tschumen*, as I have seen it spelt, is of course a Russian word. The Samoyed's name for his dwelling is 'mya.'

The 'mya' in which we found ourselves was built of

thirty-two poles or 'shaistoy,' as we afterwards came to call them. They were about twenty feet long, and at their base they formed a circle of which the diameter was about fifteen feet. They were brought up to a point and secured with a thong.

Round the outside was wound a covering of birch bark. This covering is made in three layers, sewed together with sinew, so that the seams overlap. The strips were about two feet six inches in width, and were in four lengths—two shorter for the lower part, so as to leave an opening for the door, and two longer to be carried round higher up to the apex. A hole was left at the top for the exit of smoke; while the door-way was closed in by a flap of reindeer skin.

The furnishing of the inside was not elaborate. In the middle, on the ground, was the fire, formed of drift-wood. Above this ran a pair of wooden bars, supported on perpendicular poles, carrying the long wooden hooks to which hung the big cooking pot called 'yud,' the kettle, and a smaller metal pan for the simmering of goose-fat. Away from the fire hung by three claws a tin bowl containing water for hand and face washing. You had no soap, but you tipped up the bowl and rubbed your face with water. You might do this tipping for yourself or not as you pleased. Ordinarily it was an act of the women's ministration, as it is in all the Russian cottages.

They had two very low wooden tables, a flour barrel,

a sack or two, and piles of reindeer skins. This was about all the furniture.

Mrs. Uano began bread-making about ten o'clock. They thought, no doubt, that I was asleep, but I was not. I had been awake at least as soon as they were, and watching every point like a terrier at a rat's hole. Only I lay very still and just looked through my eyelashes.

Tierrtso pounded up the flour, which was caked, for it had been wet. Tierrtso was about eleven years old. She had, like the others, a Russian name, Zornka, but as she was Uano's fourth and youngest daughter; they

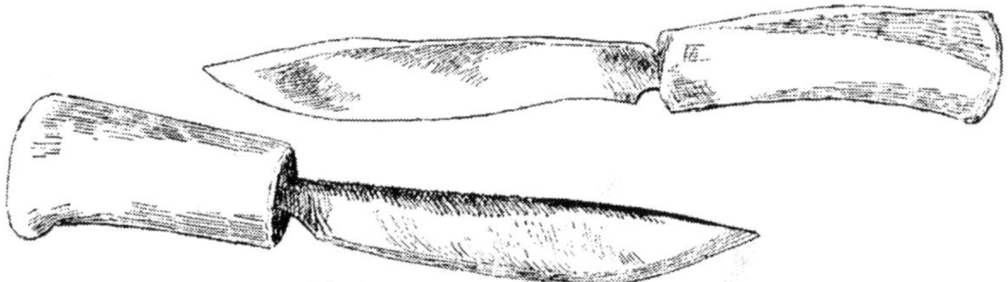

Samoyed Knives.
The lower knife is made from an old file.
Like all Samoyed knives these are chisel-edged.

always called her Tierrtso, for 'tierrt' in the Samoyed means four. The eldest daughter was married and away on the mainland; we were not to see her till much later.

Anka, the second girl, went out and cut up the drift-wood for the fire; while the eldest unmarried daughter, named Ustynia after her mother, busied herself about the fire.

Katrina, his son Philipo's wife, as Uano told us, and Niab-kutni, her little girl, did nothing on this first morning, for they were visitors at the choom. Niab-kutni's hair was short and curly. We had no other curly heads on Kolguev.

Mrs. Uano, as I must call her to distinguish her from her daughter, made bread very skilfully.

She wetted the pounded flour with water, and then

INSIDE OF CHOOM

quite spoilt its taste by mixing it with some old dough by way of leaven. Then, having made a good big pudding, with a rapid movement of her finger she ran it down both sides of a long flat stick, so that the stick carried the loaf. Finally she spun her fingers up the edges once more to thin them out and make a pretty pattern, and there you had a flat loaf an inch thick in the middle, no thickness at all at the edges, about two feet long and four or five inches across, ready for baking. Then she stuck the stick in the ground, and the loaf cooked over the fire. This bread is called 'rieska.'

While this was going on duck and goose were simmering in 'yud,' and water was boiling for Samoyed tea. This was made by mixing birds' eggs with hot water. It does not taste very well.

At eleven o'clock the remainder of the choom discovered itself. Out of burrows in the stratum of skins emerged the men, Katrina, and the baby. The baby I daresay had a name too, but they always called it Adski, for that is the equivalent for infant in the Samoyed.

We fed at noon. The Samoyeds had wooden spoons, but no forks, so they were much taken with ours. Before we could eat, these forks were handed round, critically examined, and pronounced 'good, very good.' The price was also asked, as of everything that we had. The Russians have this habit too, as every traveller in Russia knows to his cost.

The Samoyeds held the food between their teeth, and with the other end in the left hand cut quickly upwards, with the knife close to their noses. We tried this, but not successfully. It requires much practice, because of your nose.

The whole family were scrupulously clean in respect of

faces and hands. Their hands they washed often during the cooking, and always after any special office.

While the meal was preparing the men went out to collect the reindeer. This is a very interesting practice; but as I did not see it well on the first morning I will describe it in its place.

When I first entered upon this book I had thought to treat of these people in a separate part. But now I have decided rather to take them as they fall into their place in every day, by which you will not learn less about them, but learn I hope agreeably, and by an easy way.

SAMOYED WOMAN'S WORK-BAG

They had no soap, but begged a bit which we had carried for our stockings.

The whole conduct of the choom was a picture of perfect good order and kind temper. First the men fed, then the old woman with her daughter-in-law and the two eldest girls, then the children.

It is true that the dogs, of which there were ten inside, warming at the fire and picking up as they could, were somewhat of a nuisance, but on the whole the girls managed them well. And they did not interfere with Sailor.

These women, and all we saw afterwards, had a curious characteristic posture, in which they seemed as though looking for lost money. They would walk about and stand for long at a time with their elbows on their knees, which are unbent, and their shoulders lower than the level of their backs. This I found it not possible to imitate. I used to think they looked like fighting fowls, or, as Hyland better suggested, like peewits when anxious about their nests. Possibly this attitude has been derived from stooping under the rail where the pot hangs, and from the necessity of avoiding the smoke.

Now when luncheon was past we thought of travelling off. On emerging from the choom we found one sleigh ready waiting, to which were harnessed five fine deer. It was a very small light sleigh, called 'adliurs,' to distinguish it from 'han,' the ordinary travelling sleigh. 'Adliurgo' in the Samoyed means 'rapidly away,' and these small sleighs form, so to say, the Samoyed express.

On to the adliurs got Mekolka and drove off. I understood that he had gone for more deer.

Now you may wonder how I came to understand what was said to us, or to communicate in turn. I must explain that on a previous voyage I had learnt a little rough Russian, and in the meantime had improved myself by study. And although the Samoyed pronunciation was very different from the Russian, worse even than the Russian of Arctic Russia (which is saying much), still I understood some. Also, as my habit is, from the very first I began to learn the native proper names and their exact pronunciation.

Samoyed is an exceedingly difficult language to pronounce. Much of it consists of breathings, inflexions, and unlooked-for accents, which by no system of spelling can be accurately written down. But assisted by close application, and a natural ear for sound, I succeeded in catching the exact ring of the common words and sentences, so that I had but small difficulty in exchanging ideas. But now at the beginning it was a funny conversation, largely made up of signs and illustrations. Uano was a capital actor, and with this and my pencil and paper we got along fairly well. It is wonderful with how few words you can do when put to it. Thus,

Ahnglia oleynia? Nieto?
Ahnglia Tü? Yangho?
England reindeer? No?

express the question as well as half a hundred words.

While Mekolka was gone, the reindeer were brought in, and after an hour Mekolka appeared, and with him a very ugly sad-looking man, whom they called Onaska. He drove a team of white reindeer, and was evidently a man of weight. We left. We had five sleighs

and twenty-three reindeer. Uano drove five, and to the back of his sleigh were fastened four reindeer who pulled

THE PROPHET'S TEAM AT REST

Face page 142

me on a sleigh. Hyland in the same way was part of a compound organism; he was driven by Shabla, a dirty new arrival, while Onaska drove his team of five whites.

Had it not been for the Governor's letter we never should have started at all. For the reindeer were in full moult, and therefore at their worst. I insisted that we should go straight to the Gusina, and that we must be back in three days. Uano seemed to think he could do it.

So first we went north-east. But we had not been travelling more than two hours before we came to a frightful country; it was all of chasms, sudden and very full of snow, which hung over in such a way that not even a reindeer could face it.

So Onaska (the Prophet, as I called him, because of his white deer, and because he pronounced on all difficult points) went on alone to see what he could see. Returning presently he simply said 'Impossible.' Whereupon they all gaily determined that they must go back. So I came in. 'Go back if you dare; I will put the Governor on to you, and off you go to Siberia, as sure as eggs is eggs.'

With that they would see the papers again. So the papers were spread out on the ground, and they all lay down, with their noses over the paper, while Onaska, who alone could read, spelt it slowly through.

'Kolguev not here, only Petchora,' says Onaska, tapping the paper.

'Petchora or no, you go on,' say I. 'Kolguev is under the Governor of Archangel.'

So they were very sulky, but had to go; and as we couldn't go north-west, we went south-west,—a very sensible compromise. The country was much easier. We were indeed soon on the bogs, skirting those very hills along which we had come.

Four hours of this brought us suddenly upon a second lot of Samoyeds, who had but just arrived at their halting place, and were pitching chooms.

The head of these was one On Tipa. I write the word as it sounds; but I have reasons for believing that this word is nothing but Antipas. And other names I could afterwards trace to Bible origin. Onaska is a diminutive or pet name for Anania, or as we say, Ananias —Mekolka for Nicholai or Nicholas. The Samoyeds have also a practice of adding 'o'; thus Philip becomes Philipo, while Uano is Ivan.

On Tipa was a fine-looking old man, with long grey hair and a general resemblance to Moses in the church windows. He was chocolate-coloured, whereas Uano was yellow.

He gave us real tea, and we spread reindeer fat on bread for butter. I was much struck by his long train of sleighs, some of which carried boats. Afterwards I came to know On Tipa for the richest and most trustworthy man on the whole island. His district was the head-waters of the Pesanka, and he told me that he had been shooting near the Gusina that day when we had heard the shots.

I tried hard to explain that the amount of our belongings would demand more sleighs. When therefore we left at seven o'clock On Tipa and his son came with us with sleighs that brought the strength of our reindeer up to forty-six.

PREPARING TO START
In the distance a choom is seen packed on a baggage-sleigh [arki-han]

Presently we struck more into the hills, and soon met with some of our old friends the gullies.

I wish I could give even an approximate idea of the wonders these reindeer performed. Once we came to a place where was a ravine, on the opposite side of which rose a snow wall, some six or seven feet high. We were obliged to take it at this point because beyond the hill dropped down in such a manner that the sleighs could not have gone.

We got out and walked. A team of reindeer, five abreast, were brought up to the edge of the ravine, till they all stood level, breasting the long driving-pole held across them all. Then the pole was removed, and at a word all five bounded together, cleared the ravine, and though they could not take the wall they landed against the side of it, and scrambled up, dragging after them the swinging sleigh. All the teams crossed in this way.

Reindeer are enabled to do this by means of the long and sharp 'false hoofs,' which point downwards into the snow.

We slid down the hill, and crossed lower down. Sometimes we sat tight, and flew the ravines. The deer were put into a gallop before they leapt, and then over we went, sleigh and all.

We found it exceedingly cold, travelling in this way without greatcoats, for a cutting wind blew from the north, and the fog came on. But the Samoyeds very kindly lent us each a 'sowuk,' as they call it. This garment is of course no other than the north Russian 'sovik.' It is made of reindeer skin, with the fur outside, and a hood attached. We found them most warm and comfortable, but could only hear badly in them.

It is not easy just at first to sit on a Samoyed sleigh. There is only one way in which it is really comfortable. You must sit as the Samoyeds do.

I have seldom seen a picture which correctly gives this. In Nordenskiöld's Vega picture the sleigh is wrong, the deer are incorrectly harnessed, and the man sits on the wrong side of the sleigh. I fancy this picture must have been inspired by an old print which I have seen.

SAMOYED IN THE SOVIK

A Samoyed always sits on the *left* side of the sleigh, and his long rein is attached to the left-hand reindeer. But I will leave this for the present.

July 1*st.*—At 2.30 A.M. we came to a place by a lake, where was good pasturage. Here we halted, all the reindeer but five were turned loose, and the Samoyeds all lay down to sleep in the same attitude. They lay flat on their backs, with their hands, as I thought then, stretched out by their sides. But in reality only their stiff sleeves so remained, for the hands were inside on the chest.

Then I walked down to the lake, which was evidently much frequented by geese. There I waited for an hour and a half, but no geese came. There were several pairs of red-necked phalaropes flying and swimming about, and calling shrilly 'tweet, tweet.' I could not find a nest.

Returning at 4 A.M. I lay down to sleep in a hollow of a hillside. Very comfortably indeed I slept, for my sovik kept me warm in spite of the fog, frost, and a constant shower of little chips of hard ice.

About nine o'clock it cleared up, and I cooked some Liebig for Hyland and myself. The men ate food which they had brought with them.

There was a good deal of delay in getting off, because the reindeer had to be first brought in, and they had wandered far. My telescope came in very usefully here. For the Samoyeds had not the least idea where the deer were. Shabla went off with his adliurs in quite the wrong direction, and after scouring the country for some time, came back rather glum. But meanwhile I had been up to the top of a hill, and after a good look round, spied the heads of two deer a long way off, for the main body were down in a hollow. So I put Shabla on the scent, and very soon he brought them in.

In our care to avoid the more difficult parts of the hills we had dropped a good deal to the south of our line, so now we turned nearly due north, crossing again the hills we had traversed on foot, till we reached a point some five miles from the northern coast, and then turned west. Keeping this course we reached at 3.30 P.M. the mouth of the Gusina.

I was much pleased to see through the glass our tent, in apparently the same condition as that in which we had left it.

And now the Samoyeds were very funny. I pointed out to them the little green speck across the river's mouth. It was 'my choom,' I said—'mein mya.' They roared with laughter, it was so good a joke. But all the time they were looking out across the sea; first one and then another pointing at objects in the ice, as though at a discovery; but always it ended with a shake of the head, and 'yangho'—'no.' At last they all drew up together, looking very blank.

What was the matter? I asked them. And Onaska the Prophet, sulky as a cornered bear, waved his hand across the ice pack, and simply said, 'No Governor, no boat.'

So it was out.

They had made up their poor confused old minds that His Excellency the Governor of Archangel had lost his ship here in the ice, had sent us on, and was himself waiting, atop of his belongings, for relief. Or that if we were the only survivors, still there would be a fine old wreck for them to overhaul, with plenty of food and stuffs.

My conscience was clear. I had not taken them in; I had told them nothing of all this. But they were mad, very mad; and more confused than ever. For the life of them they could not understand how we came to be there at all.

They found a ford, of which we had known nothing, about a mile below our camp.

So we went across the river, standing on our sleighs, with the water all about our feet.

On the farther bank we stopped the sleighs, for Hyland and myself would go and get our boots.

All was as we left it; indeed we might never have been away. The male peregrine was still flying round the place, which much surprised me, for I judged that, as is the peregrine's wont, he would long ere this have been away to find another mate. And the goose left her nest as we came up. Her four eggs were doubtless incubated for one which I broke, as we were hungry, was in that state.

Then we pulled out our boots, all right, but earthy-smelling from their burial.

We tramped back, picked up the sleighs, and made down the river for the tent. All was safe.

It took us quite a long time to pack and stow the things upon the sleighs. And while we were thus engaged the men were hard at work in consultation about us. This delighted me very much. I could not follow them, of course, as they jabbered on, but I could catch much of the drift of it from their gestures, pointings, and the like.

We were mysteries to them, I made no doubt. No spirits, however, but good, substantial mysteries, for we gave them food to eat.

They were in a great hurry to be off; but I had made up my mind that we had earned some food and rest. And when they said there was nothing there for the reindeer to eat I bid them turn the reindeer loose, who soon would find it. So, in spite of their grumbling, they tied up five reindeer for the morrow's adliurs, and set the others free.

The excitement when we produced our English axe from its hole in the bank was beyond all holding. They all wanted it—'my axe now, yes yes, very good, make sleigh very good—not a Kolguev axe—Kolguev axe no good—how much skins for it?' and so on, and so on. But I put them off. Only Shabla, the dirty driver, watched the axe with an eye so cunning, I mistrusted him much, and you may be sure watched him too. We had two big lumps of salt pork from the ship, and on this we fed them, boiling it in our only pot. For we had made a splendid fire from the drift-wood lying round.

They did not eat much of this pork; had it been goose or reindeer they would have eaten four times as much. They could not make out what it was, and I did not remember the Russian word for it, though 'okorok' I knew was ham; so I told them it was 'okorok.' But this did not help, so pencil and pantomine as usual were my resort. I drew them a pig, but they took it for some form of seal, and said its legs were wrongly done. This put me to squeaking and grunting on all-fours, at which they laughed much, and for the greater part were none the wiser. Only the Prophet said, 'Drem, drem' ('yes, yes'); he had seen it in Archangel. Onaska, then, was not a Kolguev man, and thus I learned why he

SHABLA SMOKING HYLAND'S PIPE

In his right hand is his cow's horn snuff-box. In the distance a man goes off with the di-zha to take a deer

was master of reading and writing, as the others were not.

After this we all—Uano, Onaska, Shabla, On Tipa and On Tipa's son, Hyland and myself,—seven men in all, crowded in my little tent (seven feet by seven), where we made the evening go very merrily; for I gave them a little whisky apiece, and lent them a tobacco pipe, which went the round, though Onaska would not smoke at all, so Uano had a double share.

Then I brought out a large coloured sheet of Scandinavian birds, and also my bird-book, at which they displayed the greatest interest. They told me which were Kolguev birds, and which were not, and gave me all the native names, which will come in later in this book.

They were most intelligent about it, and had grand arguments over certain cases. I was quite surprised at their accuracy and skill, and almost everything I doubted at the time I afterwards found borne out by facts.

They said that the snowy owl nested not on Kolguev, though it crossed over the sea in the summer; that the ivory gull was common there in the winter, with many facts of that kind. Only they all insisted that the swallow was a bird of their island, but I feel sure they took it for a skua. They would give it a name, and then imitate exactly the call of the bird, to show that the name was only what it said itself.

One instance struck me much. The Samoyed name for snowy owl is 'hei-nib-chur.' But 'heinibchur' is also their name for sneeze and snuff, and when I found out later that the Russians call this owl 'Sowah,' *i.e.* snuff, I saw that it had reference to the bird's habits of sneezing, as one might call the hissing breathing noise it makes.

July 2nd.—We slept a bit, and at midnight turned out, caught up the reindeer, stowed the tent, made a fine fire, and cooked breakfast, and by 4 A.M. were ready to start.

Then I wrote a note to Powys, telling him of our change of plan, and that we were going to Scharok. I put it with the former note, and buried the tin again below Father Nicholas' cross. The Samoyeds could not make much of this ceremony, but the Prophet read the inscription aloud, at which they all looked grave and pondering, but had to give it up, and grunting came away.

CHAPTER II

OUR RETURN

It was a most unpleasant morning; a biting north wind blew, snow was falling fast, and soon covered the ground; but as it made the travelling much better, we did not complain.

We took this time a more northerly route, crossing the upper waters of the Konkina and Veliki rivers.

So well did we travel that by 8 A.M. we had done nearly thirty versts, and then we pitched the tent.

We had a most exciting and delightful drive. This side of the mountains did not seem to me to be so much cut up with gullies as the other. But it is hard to say; for the Samoyeds, who knew all the passes, often made so that we should take at any easy slope hills which would have been very fierce and formidable had we been hurrying on foot.

And very often we were able to travel for long distances along the side of the snow gullies when they were not very steep. But this was rather alarming to a beginner; for then the men would put the reindeer to a gallop, and the snow here lying not flat, but at an angle, the second sleighs, those on which Hyland and myself were seated, would slide away sideways down the decline, until it really seemed that we must be shot or dropped into the stream of the ravine itself. And the heavy-laden sleighs were often in difficulties, especially that which carried our big barrel. However, all ended well.

We slept soundly enough till a little past mid-day, and though we would have had the tent larger had that been possible, we were a little better used to our company.

We had tea with ham and figs, and again gave the Samoyeds pork, which they seemed to like better than before.

We found them very averse to making a fire; indeed they said that as they had no wood it could not be done, and that we must hurry on. But I started picking moss, and Hyland did the same. And when they saw our game, and that a fire we would have, they made a virtue of necessity, and soon had brought us much moss and creeping birch.

I let them have the lighting of it, for I hoped that we should see some curious way of obtaining fire or sparks, but was disappointed; for out of their clothes they brought Russian matches, and used them.

Long before we left Kolguev we had to all intents and purposes become, in the matter of patience and the readiness to delay, three-parts Samoyed. But in these first days of our pilgrimage we found them very trying, so that I chafed much and frightened them unnecessarily; for you can no more change the nature of a Samoyed than that of any other man, though you rave never so much. To-day you may gain your point and will, but to-morrow you shall have it all over again; so that in the end you are no better off, only more exhausted.

This is the particular way of these people to which I now refer.

The reindeer are feeding, scattered far away; but they will not make a move to bring them in until the tent or choom is struck, and every article packed on the sleighs. This done, the whipper-in, who rides on the adliurs, goes out for the deer. It sometimes takes an hour—it took an hour this time—to find them and bring them in.

Then they have to be corralled, sorted out and harnessed up, and then at last—no, even yet we do not start.

First of all they all sit down on the ground for twenty minutes or so, and snuff. Snuff-taking is a considerable feature in the life of a Samoyed. The snuff they buy from the Russians, or sometimes the vile tobacco, called 'mahorka,' which they pound up to that end. And their snuff-boxes are very ingeniously made. Usually they are of a cow's horn, the larger end filled with a plug, and the tip bored for the exit of the snuff, which is shaken out on to the thumb-nail. Sometimes they contrive a snuff-box out of wood or metal, but even then it still takes the shape of the cow's horn; agreeably to a principle well known to anthropologists, which comes out in the gourd-like water-bottle or the buttons on the back of our coats.

However, we did get off at last, and travelled very well. At ten at night they had a consultation about the tired reindeer, and it ended in a turn to the south,

TAKING SNUFF

through a pass in the hills. When we halted On Tipa's contingent left and took many of the reindeer off. It appeared that since we met On Tipa, his choom had been moved up into the hills, and was now some five versts away.

July 3rd.—We made ourselves some tea, and waited for On Tipa. His son came back without him before very long, and brought many fresh deer.

About one in the morning we got off again, and by three o'clock had reached Uano's choom.

It was so very cold and uncomfortable, for everything

was frozen hard, that I would not attempt to unpack at present. Thus for the second time we slept in the choom.

We were not so tired now, and noticed the discomforts more. I was, I must confess, not well content; for the edges of the choom were many inches from the ground, nor could we find any form of packing to keep out a cutting wind; for while there were women and children on the other side of the choom, I could not put in any claim to the reindeer skins, which were bundled there. The skins, too, were reduced in number, for some we had taken with us to sit on in the sleighs, and these were now wet or frozen from the passage of the streams.

But sleep at last we did; and the last thing I saw was old Sailor creeping round in a half-guilty way, to find himself a comfortable position close to Hyland's head.

CHAPTER III

NOMADS

July 3rd.—And now I will go on from day to day recounting events as they came about, without further digressions than may be expedient for the making clear this point or that point which does not explain itself. By these means I judge I shall best discover to you the life of these solitary people, to whom we owe so much, not only of our entertainment, but perhaps of well-being itself.

We were for a time their companions, a part of the equipage of their daily concerns, living with them side by side, feeding from their pot—from 'yud'—moving with them as they moved; in a word, we were to be Samoyeds.

At two in the afternoon of this day, Tuesday, the 3rd of July, we rose, and by half-past three the mid-day meal was ready cooked and eaten.

The Samoyed has three proper meals in the day, as we have ourselves. The word for a meal is 'zeindow.' Of these they have—

1.	Sienitz-zeindow	= the morning meal.
	Hov-zeindow	= do.
	Ortow	= breakfast.

This meal we commonly spoke of as 'ortow,' but 'sienitz' and 'hoom' both mean morning, and possibly 'hov' is simply euphonic for 'hoom.'

2. Yaliernia zeindow = midday meal.

3. Vayosoom zeindow = supper.

We had 'yaliernia zeindow' now. An excellent goose stew the women made, and when we had finished all the meat out of 'yud,' barley flour was thrown in, and a capital soup resulted. This went either by the Russian name of 'schtee,' or the Samoyed word 'yah.'

The men drank this soup in the Russian peasant way, four spoons to one bowl, but we had our iron cups. The stewed goose fished out with a bent nail was either spread on a low wooden stool, or into wooden bowls, with which the choom was provided.

The dogs at meal-times were not so bad, but outside were an intolerable nuisance. It was impossible to move without a train of yelling curs at your heels. In particular, there was an exceedingly handsome dog, an especial favourite with Mrs. Uano, Huilka by name, and he bit Hyland. He made no sound when attacking, only came quietly up and bit. Fortunately the wound was very slight, but it naturally annoyed Hyland very much.

Such a miserable day it was, however, with its snow and keen north wind, that we did nothing more than pitch our tent. This was a matter of some difficulty. It was hard to find a spot where the pegs would hold, for all the ground was sandy. The wind, too, was much against us—it was almost as great a task as the taking in of a sail in a gale of wind. But the Samoyeds helped us much, and were very envious of our tent, which they said was so much better than a 'mya.' The time came when we would willingly have exchanged with them. So far, however, we had no rain, and therefore felt full of trust in the green 'waterproof' canvas.

OUR TENT AND UANO'S CHOOM

July 4th.—This day to begin with was very fine, and we were glad to hear that our friends meant moving down to a position near the mouth of the big Pesanka river, so that they might be well placed for observing the geese which frequent the river's mouth.

I will describe the details somewhat particularly, and will say once for all in the light of later events that what is true of these Kolguev Samoyeds is true, in my experience, of their race on the mainland at least as far east as the Petchora River.

A Samoyed has three distinct forms of sleighs, the 'han,' the 'arki-han' ('arki' means big), and the 'adliurs.'

The 'han' is the ordinary passenger sleigh; upon it two can ride at a pinch. It has four, occasionally five, supports on either side, set slantingly. The 'arki-han' is really a baggage sleigh; it is very stoutly built, and has two supports only on either side, and they are set perpendicularly. The adliurs I have spoken of before; it has four slanting supports on either side.

The packing of the sleighs was done with that expedition which one would expect from a nomad people, constantly on the move because of their reindeer. Can you imagine what this means—the discomfort, the unrest of it? They can stay in one spot only just so long as the deer find plenty to eat. But a large herd of deer soon clear a district, and every day they feed farther and farther from home, and it becomes more difficult to bring them in. So after a day or two there is nothing for it but to pack up all your belongings, and move away to new pastures.

Small wonder, then, that these people were good packers. They were. Everything fell into its place as if by magic.

The women and children did all the domestic part. They took down the chooms, and packed them on the 'arki-han,' and the pots and skins they stowed also. Meantime the men got barrels, the boat, and the firewood on to the sleigh.

These boats ('ar-noh' in Samoyed) are very light. They have no keel, but usually two bilge keels or runners on either side, so that they are pretty steady, and run easily over the mudflats. They are nine feet long, with a three and a half feet beam.

You are to understand that the chooms were placed on a circular sandy plateau, of which there are many on the tundra ; and that such places are always chosen, when possible, for the following reasons. There is a good view from them, so that the reindeer can be kept much in sight; they are dry and healthy; and from the hardness of the ground are well adapted to the practice of such exercises as I shall describe in their place.

Now all was ready for the deer. The team of five fast and strong bucks which had been staked out by the choom all night was harnessed to an adliurs. One such team is always kept up in this way in order that the whipper-in may have them ready to his hand.

Mekolka is to take this team and bring the deer up. Off he goes at score, followed by some of the dogs. But I have never introduced you to the dogs.

Latori was king of the pack. He was an old white dog, with one eye only, and the sweetest disposition you

IN LINE

Face page 165

ever met with. He was respected as a patriarch by all the rest, and retained his throne by moral influence; for if might were right, as among the dogs of the Eskimo, the rule would already have passed to Ni-arr-way.

Ni-arr-way was a fine black dog, the biggest of them. When Ni-arr-way gets to be Latori's age he will be just such another in character; for even now he is the soul of kindness, and a remarkably useful dog—as you shall see.

Thirdly there came Huilka, the handsome treacherous one. Red he was, with a lovely fox-like coat. Then Pasco, who had the puppies later on. Mandaluk, a lady also, and very pretty but shy. Sixth, Sierko, about ten months old; white, ragged, and always dirty, who barked incessantly at nothing at all, and I think was touched in the head. Hurpey, a piebald youngster, who promised better things. Mno-coh, a dog like a timber wolf, whom I seem to know little about. There are dogs, you know, with whom you may live for a long while, and yet notice very little; independent characters, who 'gang their ain gate.' Mno-coh was one of these. There were also puppies which came at intervals, and were loved and nursed by the children as they are at home. And sometimes they disappeared mysteriously, as they also do with us. But of those who survived our favourite was little Chua. Chua would play with you by the hour if you gave him a chance. He was one of those electric pups you sometimes meet. If he was asleep at the other side of the choom you had but to snap your fingers and whisper 'Chua,' and in a moment he was sprawling all over you, wriggling like an eel.

These, then, were some of the dogs.

Off went Mekolka on his sleigh, and rounded up the the herd. It was not so easy as it may seem. There were five hundred reindeer in this herd, and they were scattered far away. But the team was taken at a rapid trot round and round the herd, and the dogs worked also, as sheep-dogs work at home. The bulls and steers would always run from the dogs, but a cow, especially if she had a youngish calf, would often charge and charge again. But the whole herd is brought up at last; the steers and bulls being then penned—for in a strong herd such as this the cows are not worked at this time of year.

As the deer come up all the 'choom' turns out to meet them. The draught deer as a rule are pretty quiet, and draw up together more or less of their own accord. They are brought up on to the level rise on which stands the choom, and then the women and children encircle the lots, with a bit of light string held in the hand, while the men going into the ring make up their various teams. Some of the deer have to be caught and haltered; some are so quiet that an arm over their necks is sufficient to guide them about.

Then the creatures are harnessed up, and off we go.

The two little girls each drove their team of three reindeer, and wonderfully well they managed, jumping on

REINDEER COWS AND CALVES TRAVELLING

Face page 167

to the sleigh when the deer were at the trot as easily as their elders.

The light-heartedness of these children was delightful to see. They wanted to drive races, but were not allowed, because the deer were not at their best. But they were laughing and chattering all the time, and were much concerned about their puppies, which kept rolling helplessly about on the sleighs.

All other movable things were roped down on the sleigh with strips of sealskin; but you can't rope down a fat, round puppy.

We were not yet promoted to team-driving, so I had nothing to do but sit on my sleigh and take snap-shots with the camera at anything I chose. Mr. Nettleship's picture gives an admirable idea of the general appearance of a travelling herd of cows and calves.

Mekolka was well occupied. With his light adliurs and his fine team of five he was continually scouring round at the gallop or hard trot, bringing up the stragglers.

When we reached the Pesanka we found it pretty much swollen because of the melting snow, and it was some time before we could find a fordable place.

First of all the cows and calves were driven across. The dogs hunted them over. The main body went readily enough, but some of the cows whose calves were small would now and then rebel, calling loudly, and charging back on the dogs. And one or two on finding that their calves had remained behind on the bank turned, and came pounding back, swimming and dashing across the shallows, and taking no more notice of the dogs than if they had been stones. So they rejoined the calves, and then both crossed together. The little things swam admirably, carrying their heads well laid back, just as their parents did.

It had been a most picturesque and charming sight. I sat and sketched it while the Samoyeds, in default of

A HALT

other occupation, squatted down close to me, rather too close to be quite pleasant, and looked on.

Of course at this place was the regular ford, but the water was unusually high. It was quite impossible for us to cross straight over; so now Ni-arr-way's part came in, and a very proud dog he was. He had to find out for us a passage by the shallows. For, as Hyland and myself had by this time every reason to know, in these rivers a passage may often be found, though you take long about it. You may have to grope halfway across, and then perhaps a quarter of a mile or so right up the middle of the stream, and possibly at an angle down again, but as often as not you come out all right at last.

And Ni-arr-way did this for us. Into the water he went, and puzzled out the track. He might not swim; that was forbidden. Whenever he came to a deep place he turned, and tried another tack. Close behind him followed Mekolka, standing up on a pilot sleigh. They were a long time about it, but at last they emerged on the farther side some three hundred yards above.

Meantime the women had been gnawing bones and horns, and the children running about picking dead grasses to put in their shoes. This grass they call 'sti-el-ka,'—a pretty name.

Now it was our turn to cross. We went over easily and well. As Uano and I were the first over, I was able to sit and watch the long train of sleighs as it wound its way across. Twenty sleighs mean some eighty reindeer, so it took some time for all to cross. And what do you suppose happened then? Why this. A wretched white puppy, quite big enough to swim if he had had an ounce of pluck, was found to be running up and down on the farther bank, and squeaking miserably. 'Tasso' is 'puppy' in Samoyed. But not a soul called 'Tasso,' only Mekolka turned again and drove all the way across for the creature; and then, with the perverseness of puppyhood, this little stupid would not trust his deliverer, but wriggled away out of reach, careered around, and took as much catching as an autumn grasshopper.

At last, when things seemed at the worst, he suddenly rolled over on his back, smit by a sense of his own impossibility, in the ridiculous way that puppies have, and was carried off in triumph. I timed this sleigh recrossing; it took twelve minutes. Soon after we rose a little hill, and pitched. Then I went out with Hyland for a stroll.

We were walking about thirty yards apart, along a high peaty ridge, when Hyland called out, 'What's this bird, sir?' It was a dotterel, and was sitting quietly close in front of him. Whether it had a nest or no we could not determine. It flew off with no deceptions.

Then we circled round, and presently a second dotterel ran away, evidently off her nest, for she was shamming lame. We had no difficulty whatever in finding the nest. It contained four eggs. The dotterel which nests with us, though now few and locally, is a bird of very beautiful plumage, and the female is more strongly coloured than the male. It has for trivial name the '*foolish*' dotterel, because once it was so easily taken in the nets.

We did not touch the nest, as I wanted to see something more of the birds, but came in and had a capital meal of goose soup, black bread or 'rieska,' and Samoyed 'tea.' This is easily made by mixing up raw eggs with hot water. I cannot recommend it on other grounds. It is not equal to the best China tea.

By the Pesanka to-day I found a charming little pink primula, and several other plants which will be found at the end.

The Samoyeds were much mystified at the trouble I took about these plants, not being able to imagine to what use I would put them. Although they have distinct names for many of the birds, the plants, except the edible forms, are all lumped together as 'omdoh,' which is to say, 'a flower.'

July 5th.—During the night a big wind arose. It grew and grew until it raged all day, a gale with rain and snow. A vile day; you could not walk; to go out was to be almost blinded.

And now, for the first time, I lost faith in the tent. It was one of those known as 'Whymper' tents, which are contrived for Alpine work. They are made of 'Willesden' canvas, recommended as waterproof. These tents are excellent in dry and still weather, but in a wind the sides bulged in badly. I always thought that this might be met by a portable set of ribs; but how they could be fortified against rain with wind I do not know. Rain which falls quietly runs off well enough, but driving rain comes straight through. Very soon, then, the tent was flooded; and for the rest of the day we were fighting the water, making what contrivances we might for guiding runnels clear of our things. By putting blocks under the floor we just managed to keep a corner dry where we would sleep.

Not doing this, we were skinning and blowing eggs; and it was a great disappointment to find that two grey plovers' eggs had broken in the botanical case where I had packed them when the bacon was done.

We were not more than ten yards from the 'mya,' so I could keep an eye on anything of interest there. Mekolka, one of the most ingenious fellows I have ever come across, was making a ring for his finger. Round a stick he cut a groove, fitted a bit of paper over it, and then ran in some white-looking metal which he had heated on the fire in a hollowed piece of wood. What this metal is I do not know; it is not lead, but evidently some alloy which melts at a low temperature. The ring had a fine big stop like a signet-ring. The wood was cut away, the ring removed, finished with the knife and with sand. He went on working until he had a fine heavy metal ring, as well and accurately shaped and as well finished as any plain gold signet-ring which you can see.

The Samoyeds are very clever at this work. Their chisel-edged knives, of which I have told, have handles ornamented with metal by the same plan. Some of these patterns are quite intricate in character, and we never saw two alike. Every man, woman, and child carries one of these knives. We had not a single case of accident from them while we were on the island. I wonder how many English mothers would like to trust their children of five and six with long, sharp knives like these. But the use is, with the Samoyeds, a case of heredity. They use the axe, too, wonderfully. A little girl of ten or so—little Zornka, for example—would cut up the wood for the fire with great precision and speed.

July 6th.—You are to understand that we made our hours for meals by the sun. At present it was about like this. At ten or eleven in the morning the Samoyeds rose, and the men went out to tend the reindeer and to settle matters about the camp. There was always plenty

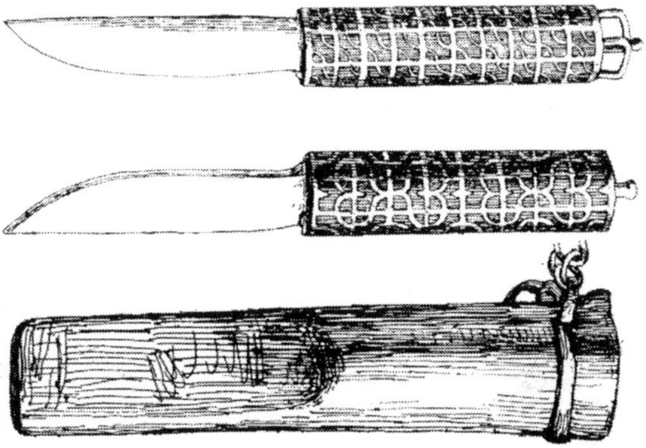

to be done—sleighs to be made or mended, harness put right, reindeer brought up and inspected. Meantime the women were getting 'ortow' ready. They fetched for water melting snow in wooden tubs; they made the fire, plucked the geese, and finally called us to a well-cooked meal at noon.

At seven in the evening 'yaliernia zeindow' fell, and 'vayosoom zeindow' at midnight. After the morning review of the reindeer I was not a little surprised to find that many teams had not been set free, but remained tied up to the sleighs. Whereupon Uano told me that the 'choom'[1] to-day was moving on.

This would have been greatly against my own idea, for we were now in a beautiful little spot, and one that promised well for the birds. For at the foot of our hill flowed the Pesanka, which we could trace very far away as it wound across the tundra. More than one little lake we could also see. Nor had I forgotten the dotterel's nest.

Our property was still wet from yesterday's soaking, and must be dried; and between these I made out sufficient reason why we would wait here another day.

I do not say that our host agreed willingly to this; on the contrary, it raised a universal storm of protest in the camp. But I had always one and an invincible argument on my side. This was nothing else than complete indifference to anything which they might say.

Of course they might have driven off and left us, but so early in our acquaintance they were too greatly afraid to do this—just as afterwards they developed such an affection for us as would have kept them with us against our wills.

It was blowing about half a gale from the north-west, which dried up our belongings by about three in the afternoon. Then we went out, and again visited the dotterel's nest.

[1] The word 'choom' is used not only of the wigwam-like dwelling, but, as here, in the sense of the family or household, as we say 'my house.'

On one small lake were three black-throated divers, of which we shot one. Sailor retrieved it, though it pecked viciously at him all the time. While we were standing there and shooting, two more of this species flew over and dropped into the pool, which was not more than sixty yards across. Each time that one of these birds came to the top it barked and yelped so exactly like a dog that at first I was completely taken in, believing that Sailor suffered correction at Hyland's hands for doing something wrong.

This lake was very pretty. There was a steep bank on the farther side which was many feet deep in snow. It hung over the water so that we could not go that way; but where we stood was only marsh, and all ablaze with marsh marigold.

To-day Uano produced three fox's skins. One pure white, one red, and one blue. They had all been taken in the winter-time.

In the evening we played some games of draughts with old Uano.

I had noticed that the table off which we fed was scored with cross-lines, and worked in a chequered pattern, and this evening Uano explained that it was for the game 'hardranger,' which after all was only draughts, with variations. They used for the pieces the Russian words 'kiesler' and 'dama,' and from the Russians perhaps they had the game. For draughtsmen we used little stones, which came out of a skin bag.

Uano beat me game after game, roaring with laughter at my defeat. Old Ustynia looked on with her wrinkled face and kindly twinkling eyes, taking the keenest interest in her lord's success. Hyland played it better than I, but he was worsted too.

July 7th.—It was a lovely sunny day, though the wind still blew keenly from the northward.

Always most anxious to keep any undertakings made with these people, so that they might come to rely upon my word, I used every effort to have all ready and stowed on the sleighs by half-past two, the hour they had fixed for moving off.

But the reindeer had strayed away, and so we all stood tentless and choomless until six o'clock. And then we moved off.

Among the herd was one little calf only born five days ago. Yet this small creature kept up most gallantly. Every now and then it dropped behind, and its mother anxiously waited, lowing, and charging the dogs, who tried to hurry it on. But just when we thought it was really so tired that we ought to lift it on to a sleigh, it pulled itself together and scampered off.

We pulled up now and then for the deer to rest, and once Mekolka left us. He had picked up by the way a reindeer's shed antler. This he bore away to a spot on the tundra, where was a circular, low tower, built entirely of reindeer's horns. On this he placed the horn he bore, but he brought away, after much selection, another in its place. The Samoyeds pile the horns in this way for storage and exchange. Under the action of the sun and the salt winds the horns are bleached and workable, and are useful then for many purposes.

We were now for the first time on a regular sleigh-route ; from continuous old traces of sleighs you could see that this was so.

But in the evening we reached a charming spot in full view of the ice of the sea. By us, on the other side, were several little pools. The place itself was a gravel, stony rise, not lofty, but high enough to command the tundra which lay around.

It was evidently an old camping-ground, and here we stayed.

SLIPPERS MADE BY SAMOYEDS

CHAPTER IV

PESANKA TO SCHAROK

July 8.—No day in any country could possibly be more lovely than this. I might have been back in the mountains of California, when the sun is despoiling the snow, and the wind comes down the valleys so lightly that it scarcely stirs a leaf of the poplars to fright the black-tailed deer. This sense was often upon me during the day. But could I paint as became the subject you would have a very different scene from that. Long descriptions tire us all, but for once I should just like to try and set you down where we were now. So will you, for one moment, shut your eyes, and give yourself to me?

Just behind us was a gorge filled with snow. Up this the sleighs had wound from the river to find the plateau where we were. But very soon on that northern side the hills dropt down, and there was the Pesanka coiling away over the plain in the sunlight like a beautiful glittering snake. And the wild geese went up along it from their feeding-grounds, and all about its edges small waders ran and fed.

Our hill sank in front to the tundra itself; very grey at

first because of the lichens, as it rolled along it changed and changed again. For it took colours from the sunlight and colours from the clouds. Here it was purple, which deepened back and deepened back till suddenly a spot lay lit to emerald, as the sun caught a moss-flat or the grasses round a tarn. And here it was blue—a blue like the haze in England when the hay is in rows in the sun. And if you sat and watched a little longer, lakes began to form from separate small centres, and widened and grew till they formed a chain of waters spanning the tundra from hand to hand. Though you knew they were but phantoms of the mirage, transformed from little drifts and cups of snow, it mattered little—you had your picture all the same.

And away to the west lay the ice barriers, dense upon the water and ridged upon the sands. Only now, lifted up in the sunlight, it was no more the level ice-pack, but noble cliffs of quivering whiteness as round some enchanted isle.

There was more than this, much more. But these are the broader touches in the scene.

This was Sunday, and that perhaps was why the women all knelt in a row by a stream and washed their hair. Then they combed it with their fingers and tied it up. It is straight and black, and it reaches nearly to the waist. Also, because it was Sunday we cooked one of our four tins of preserved beef, and stewed some apple chips.

I don't mean to imply that the Samoyed food is not

good. It is; but when you have eaten white-fronted goose, bean goose, king eider, and long-tailed duck for many days on end, you begin to feel with Mark Twain's American after watching the dead bodies, 'Cussed if I want any breakfast.'

And after all the white-fronted goose and the bean goose, though they have different names and different feathers, when they are cooked are only—goose.

Of the other two, however, king eider is excellent, but long-tailed duck very poor. We came to hate long-tailed duck.

Verrmyah, also called Philipo, Katrina's husband and Uano's eldest son, appeared to-day. It seemed he had been southward. He brought very cheering news. The ice, he said, did not touch the south of the island. I asked him whether any ships could get through to the south. He did not know; there might be a channel, and a boat *might* get into the gulf; but truly, he did not know.

UANO AND VERRMYAH HIS SON

But this set me thinking whether we ought not to be somewhere nearer the gulf or Scharok harbour in case a vessel should come. For this was just about the time when, as the governor of Archangel had said, the gunboat would be going on its first trip to Novaya Zemblya, and could set me ashore on Kolguev.

But it was pleasant here, and I was loth to go. Philipo was quite a good-looking fellow, with a frank laugh that caught me much. Athletic and well-built he was also, if you except the bowed legs which nearly all these people have. Philipo was his Russian name; his proper name was Verr-my-ah, just as Uano's was Süns-koh.

They skinned a dog to-day.

A young reindeer (not the little one) had died last night. They didn't in the least know the cause of death, and yet they let it lie about until all the dogs had gorged themselves upon it. As a consequence of this there was an epidemic of swelled throats, all the dogs looking as if they had mumps, and had them badly. And one, Huilka, died. You remember Huilka, the old women's pet, how handsome I said he was, and how he bit Hyland's leg. I never had reason to dislike him myself, but was relieved when he was taken from us, for he and Hyland had never made it up. He was a wonderful catcher of birds.

I saw to-day many red-necked phalaropes, a pair of scaups, and several long-tailed ducks in the pools. All the male long-tailed ducks had lost their tails.

July 9.—To-day, with a south-west and warm wind (the thermometer in the tent at 10 A.M. stood at 50°), the musquitoes began to be troublesome in the marshes. I found it impossible to escape them till I walked down the bed of a little stream between walls of snow. Here I sketched a very big sandstone boulder. It was quite

circular, and was just nine feet across. But it lay in three pieces, cracked right across the lines of formation by the expansive force of ice.

Then I went and lay down by a lake. Had it not been for the musquitoes it would have been most pleasant there. The black-breasted dunlins were running and flying, or sang like grasshoppers from the tops of mounds. I found four of their little ones crouching in the grass. They were all downy, but with big strong legs.

The little stints I could not understand. They were now about in lots of six, nine, or ten, chasing one another round and about, as on the Kriva all that time ago.

And I watched four pairs of red-necked phalaropes for a long time, with no result, at least towards finding their nests. They swim well, in a curious, jerky, perky way, with heads straight. They somehow recalled to me a crinolined lady by Leech, picking her way across the street in the pages of *Punch.* Mincing—that is the word—the red-necked phalarope seems to mince.

And then, while I was looking at a red-throated diver, it swam to the farther side, and then I could see it through the glass sitting in the reeds. So I made a circle, and came quietly upon it, whereupon it stretched out its long neck, and shot down a lane of shallow water that led from its nest, and away out into the lake, cutting the water for many feet, just as an old pike does when you have him by the lip.

But now I had weighed to more purpose the question of our course.

If the ice was really away from the Waskina, ought not a message to be left waiting there for Powys should he call? If Scharok harbour were open out to sea, then we should be making soundings, and finding how the channels lay against the coming in of the *Saxon* or the gunboat.

After I had finished the work with the prismatic compass, and all the entries in my route-book, which always took up much of my time, I told Uano we would move to Scharok.

We left at about four o'clock in the afternoon. For good-bye the women said 'prostee,' which is the common Samoyed formula. It may be Samoyed, but it sounds like a corruption of the Russian 'pra-schei' or 'pra-scheite.'

We crossed in the order named, the Tinyan, the Barakova and the Peinmur rivers, all running through the same flat, and then came upon a really fine lake. This Uano told us they called 'Solnoi Toh,' which is to say, Solnoi Lake. The Russians, he added, called it Solnoida or Soldonoida Lake ('ozero').

Soon after seven o'clock we made the passage of the Baroshika, the largest river, said Uano, after the Pesanka, on this coast.

An hour and a half later we rose a slight hill, and there, some four miles off, were a row of little huts, plainly visible against a background of ice. They were not much to look at, but even after these few days of our wanderings seemed about as strange as though in the middle of Gobi you should come upon a modern hotel.

Here the men pulled up the reindeer, flung down the

WOMEN HARNESSING UP

Face page 184

'toors' (driving rods), and sprang from the sleighs. 'Scharok, Scharok,' they cried together—'Sarco in the Samoyed—in the Samoyed Sarco.' They always put everything twice like this, for fear that I should miss it. For after they found that I wrote down their words, and used to try and use them, they were at the greatest pains to teach me all they could.

And then they snuffed and the reindeer fed. The ground was well lichened, and the deer were just half an hour feeding steadily before they raised their heads. Then within a minute or two of each other every head was up. But we left them no time for rumination, for again we moved off.

And then we took to the snow at the sides of the mud creeks, and then to the mud itself, on which the sleighs ran very well.

Splish, splosh went the feet of the deer, squirting up unsavoury mud which the tide had left. It was red at the top, but below this crust was lying liquid and black.

At last we came on the huts—'isba' and 'ombara'—sleeping-huts and storage-huts. All were deserted. One of them, Uano said, was his. 'Have you no boat of any kind?' I asked. 'Yes,' he said, there was a boat, but it wanted doing up. It was lying up among the sheds, but we hauled it into a handier position, nearer to the sea; for I meant to use that boat. Then I pitched the tent, and gave the men some tea, and just before midnight they picked up their toors, jumped on their sleighs, and were off at a gallop, but not before I had made Uano undertake to come for me in nine days' time, so that I might go and help to catch the geese.

Meantime we were to be alone.

PART IV

WHICH treats of our solitary life at Scharok harbour—With how we explored the district—How we sounded the harbour—Of 'the waiter' and his ways, and of new faces—A baby is described somewhat particularly, and a Samoyed Saga is given—The reader will learn with other matters how we met the father of the island—How we took the wild geese—And finally, how through want of bread we were compelled to return to life with the Samoyeds.

CHAPTER I

IN SCHAROK CAMP

So the Samoyeds were gone, and again we were alone.

There was no help for it, and the best we could do at the moment was to prowl about and see exactly how we lay.

We made for the huts—they were about as cheerful and promising as a group of tombs.

Three were 'ombara'—rickety shanties, designed as a cover for barrels and skins.

Three were 'isba,' or dwelling-huts: one was locked, the other two we entered. The first was Uano's.

The door—so low that you had to bend down as you passed—opened into an antechamber, some ten feet square, and blocked with barrels full of fat. The floor was of earth, covered with a good deep layer of melted seal-blubber. It had run from the tubs out through the doorway and was smeared all about. A still smaller door led into the living-room. On either side of this was a wooden settle, a small table stood between the two, a tiny window looked out across the sea, and on the wall hung a little Russian cross of wood.

The second hut had its complement of blubber in the entrance-room, but the inner part had been evidently used as a chapel. There were the Madonna and a saint or two, left cruelly alone. Before them hung their lamps with none to light them, and the gilt about their setting was all tarnished by the fogs. On the floor was a packet of incense, a bundle of tapers, a censer and a charcoal-heater—and this was about the sum. It was depressing.

THE CHAPEL AT SCHAROK

Outside it was more melancholy still. For here was a place of burial, but not railed off. Only there were many Russian crosses with legends and queer patterns, some fallen down.

Foxes had burrowed into the graves; glaucous gulls sat on the crosses and scarcely moved at our approach. Seal-fat was everywhere; in great lumps and smearings upon the grass, and on everything that lay about.

Reindeer skulls and bones and horns showed through the thin grass on every hand.

Never—*never* need a man feel lonely when he is away out in the wilds with only nature there, for so long as he has flowers, birds, and streams about him he has always friends at hand. But for downright, sickening, malicious melancholy put me in such a place as this, where man has had nothing but dreariness to give for all that he has taken away.

We walked farther, a little up the coast. The tide was out, and there lay open a wide expanse of ooze, cut up here and there by a creek from which came the voices of birds. The cliff was not higher than some thirty feet, and all along the coast-line ran a broad, sloping escarpment of slowly melting snow. Away out beyond the mud-flats, as far north and south as the eye could reach, was the white line of the ocean ice where it struck upon the outer sand-banks.

We came upon a neglected boat and a neglected grave. But the grave had once been above the ordinary, for it was covered by a wooden tomb. In this grave, as I afterwards learned, Uano's father was buried. He had not always been a Kolguev man, but had crossed from the Timanskii tundra in the island's palmy days. I saw that the willow-grouse had been dusting about this tomb, and when I raised it, underneath was a snow-bunting's nest. But it came on to rain pretty sharply, and so we turned back.

You quite understand where we were now? At Stanavoi Scharok, *i.e.* Scharok harbour, the one and only harbour of Kolguev. For Uano had explained to me that the Waskina was no harbour at all, that they could not even get their little boats in there, and that, though there was one river—the Pugrinoy—without a bar, still (as was clear from his description) there was not enough water in this to float the *Saxon* or any other sizable boat.

So now there seemed to be nothing for it but to wait within sight of this one entry for the ice to go.

July 10.—We found it very hard work to get wood for the fire. No drift-wood is cast up on this side of the island except on the outer rim of the sand-banks. It was impossible for us to reach these, for there lay between ourselves and them a good two miles of water and mud. But we found the foundation timbers where a hut had stood, and managed, after much labour, for it was like iron, to cut some of this with our little axe.

Hyland, who since during our stay with the Samoyeds had been very fairly well, was out of sorts to-day. He complained much of giddiness, and had to sit about. Our stock of medicine was not a large one, though, fortunately, in the end it proved enough.

Last night as the tide was ebbing, nine-and-forty long-tailed ducks came down the creek with the stream and lay just below our tent. They were still there this morning, but well out of shot.

Later on I made a plan of Scharok harbour and its entry, but it will be enough to say now that there is nowhere any expanse of deep water; only immediately under the huts runs a wide creek which, after three windings, finds it way out into the sea over a low bar some two miles off. This entry we could clearly see, because it was free from grounded ice.

There is far more to be done when settling down into camp life than many might suppose. As we had no bucket or water-can, the Samoyed plan of melting snow for drinking purposes would not do. But we found a place down by the edge of the creek where a tiny stream came under the snow. Breaking a hole through the snow-crust we tapped the water and found it good. But this spot was a long way off from our tent.

Also we collected a quantity of tightly rolled-up birch bark, which we found along high-water mark, for lighting fires.

I have often puzzled over the origin of these rolls. They are found in abundance on our own east coast after heavy northerly gales, and are spoken of by writers on Spitzbergen and Novaya Zemblya. It has been suggested that they are used as floats by the Norwegian fishermen. I scarcely think this would explain the immense quantities in which they are found. Every tide brought in a fresh supply, and we had a constant store for fire-lighting. For this purpose birch bark is superior even to pitch pine, for it catches instantaneously, and burns even in a high wind. When in need of a fire on the mainland we cut some green bark off a growing tree, and found it burned equally well.

Sailor's kennel was a question of some moment. We had not a box or barrel, nor anything that would answer. However, the old dog himself solved the case. For when we had provisionally staked him out under the lee of a sandy mound, he set so vigorously to work that very soon he had a snug retreat well out of the wind. When the wind changed he made a second on the other side. And now let the wind shift as it might, he had only to move to the one or the other to lie as snug as a fox.

'Now, Hyland,' I said, when all was complete, 'we have two things to remember: first, that we have now to trust to our guns for food; and secondly, that we can't afford to waste a shot.'

So leaving him to try and sleep off his feeling of illness, I set off walking down the coast, which at this point is much like the Kriva district. There is the old raised beach of yellow sand, with the bogs and flats beyond. But the ground—as a bird and flower ground—seemed in no way equal to the Kriva, for the flats were a monotonous waste of moss and peat, which apparently ran on and on till it rose to some high hills or mountains in the extreme distance.

I found one willow-grouse on the sand-ridge, and that I secured. It was getting its dark plumage. I saw no other bird of this species, whereas hitherto they had been round us every day.

In a pool I also shot three long-tailed ducks, so we were secure for supper.

A nest of four little stints I found this day is worthy of mention, because it was the first of those instances of young birds being kicked out of the nest by the parent of which I propose to write later on in the book. The old bird as she jumped and crept about squeaked just like a house mouse.

When we cleaned the willow-grouse for the cooking, we threw the insides on to the grass about ten yards from the tent, and these the Arctic skuas came and carried off.

The pink sea-glaux was now coming out and the pretty blue Jacob's-ladder. And everywhere cuckoo-pint was in full flower; but its flowers were white and not lilac.

On the mud-flats I noticed a new wader among the dunlins. It proved to be a Temminck's stint. The glaucous gulls, who were our very intimate friends, used to carry their bivalves from the creek away on to the swamp behind the tent. They had quite a collection of shells there.

This evening there was rain and fog, and it was pretty dismal.

The wind chopped from south-west to north-west and back again; and with every change the ice, as it separated, went off like great guns.

The ice was never still. It set up or down with each tide, and the blocks took often most fantastic shapes.

Most bitterly did I repent having come down here, which

1. JACOB'S LADDER
(*Polemonium cœruleum*)

2. MARSH SAXIFRAGE
(*Saxifraga Hirculus*)

I should never have done but for Verrmyah's story of the disappearance of the ice. For I could see we were in a bad place. And the ice? I climbed up to the roof of Uano's hut in the very early morning when the air was clear, and instead of Verrmyah's open sea this was what I saw: To the north a line of ice—how wide I could not tell—but too wide for a vessel's approach. Down the eastern side ice—ice beyond the harbour entry, beyond that an open way, and then ice again. The Gulf was just one single ice-field; and though there may have been a patch here and there of free water, only madness would try to take a vessel in.

July 11th.—With a light wind from the east and the thermometer at 50° F. in the tent at noon, it felt quite warm.

When Hyland went down for the water this morning he found a single brent goose in the harbour, which he shot at and winged. After this bird old Sailor went. The tide was running out strongly, and the old dog had a tough job against it. He stuck bravely to his bird. But it was no use; the goose had the legs of him all the way. It never dived, only it would let the dog draw up, and then, just when he was quite certain the game was his, swim quickly out of reach uttering its defiant cry—a single 'Wank!' At last I grew rather anxious about the old dog in the cold water, and called him in.

Hyland in the evening brought me in a sanderling in

summer plumage. He had found it by the harbour-side. This was in its way a find. For the sanderling, which does not nest with us (though it is believed to have bred in Iceland), is suspected of nesting on the mainland; on the Petchora, for instance, and I had great hopes of finding its nest.

But you will be tired of my hopes—they were so often disappointments.

And I do not think I will go into camp details much more.

A big bit of everyday would only read like this:—'Fetched water; cut up wood—chipping at logs of iron with a little single-handed axe; cooked, ate, went out to kill for supper; pressed flowers; blew eggs; prepared specimens; wrote; wondered about *Saxon*; watched ice.'

Somewhere against the ice the breakers hurled themselves with the noise of battle, and again the ice went off like guns. I liked this cannonading of the ice. It was a fine sound, chiming well with the wild cry of the divers and the call of the geese overhead. And when you walk along by yourself in a lonely island you—I suppose every one does—come to look on all these as your creatures or your friends. They are not a bit so, of course; just the opposite. You are probably in their view the one thing unresponsive, the one fact with which they have no relation—a thing out of all reason, a great nail-booted, shot-blazing, contra-natural blot.

But you may fool yourself—you may have your Barataria none the less. You can lie back upon the moss with the sea-wind in your face and be—greater than Sancho—be a Prospero who holds all the strings.

Your island perhaps is assailed by legions, half-Viking and half-monstrous forms; but let the assault break never so fiercely it cannot silence the guns.

Down at your feet swoops the falcon with the story of the battle, for it watched as it hung in the cloud. And the glitter-winged divers carry your commands, and large-eyed seals come up to spy in unsuspected corners, and, silent and secret, disappear again.

All the things have their places according to your fancy, and you may detail it as you please—a foolish baby game, but there is no other to play in such a lonely isle.

July 12th.—We really couldn't stand the seal-fat any longer. It was smeared over the grass and all about, so that it stuck on our boots and came into the tent. It was so universal that we could not avoid it, so adhesive that we could not get it off.

We moved the tent lower, away out of its reach, and nearer to our snow-pool. Also we collected a heap of chips, which we loaded on the floor-cloth and dragged away.

A snow-bunting lightened our labours much. For he sat on the roof of the huts and sang a lovely song, with no touch in it of a bunting's scrape, but a true wild song, and very like a chaffinch's in spring.

A yellow pyrethrum came in flower and a white lychnis.

We walked down the coast. We found a little stint's nest with four eggs, and the old bird—a male—performed the most astonishing antics. He looked, as Hyland said, for all the world 'like a dancing doll.' Of four young red-necked phalaropes two were kicked spinning out of the nest by the mother. Two young turnstones bolted out of their nest and ran away.

We had all day a very strong wind which was a gale at midnight, with fog and so cloudy a sky that I lit a candle, or I could not have written my notes.

An hour later I looked out for a moment before lying down; the sun was out, and there, right across the ice-white sea, was the most splendid rainbow I had ever seen.

July 13th.—In the early morning such a stiff gale blew that I feared indeed our tent would have fallen; however, nothing gave. Poor Hyland complained bitterly of cold hands and feet; so I let him lie.

It was impossible to make a fire in the usual place; but a little out of the wind, in the water gully, I contrived after much patience to cook a piece of bacon.

After we had eaten this I went for several hours' walk alongside of an inland creek which lies to the north of the camp. I suppose I saw twenty glaucous gulls, but with only two immature birds between them. This made it pretty clear that their nests had been robbed. Many Siberian herring-gulls were also about the creek, and of these many were young birds.

At half-past three that afternoon Hyland was so much better that we walked down the coast to inspect the sand-banks.

I had had some hope that we should reach the gulf, but, owing to the great difficulty in crossing many inlets, we were obliged to turn homewards when we had gone five miles.

I took a grey plover's nest, and removed, with its surroundings, one of the little stint's. We also saw two lots of baby little stints.

I shot three long-tailed ducks, and one of them when winged gave Sailor much sport in the retrieving. And here I may notice the different methods pursued by dogs in this attempt. Sailor had no other idea but to swim about, waiting for the re-appearance of the diving bird, and making then the best shot he could. So the dog was often head under water, and often foiled.

But a Samoyed dog who went to fetch a winged eider, as we rode to the Gusina, had behaved with far greater address. For he swam quietly and warily, always trying to come on the duck from behind. And so well succeeded that on the third or fourth attempt he was able to take the bird quietly in his mouth before it was at all aware that he was near.

After this, leaving Hyland to lie down and rest, I pushed on across some small inlets to find out the nature of a heap of wood piled at the point of a spit of sand.

And this it seemed served two purposes. It was both a guide or beacon, and also a means of supplying the chance wanderer with a log or two for the fire. For this sand-spit was quite evidently a leading passage by which the marsh could be entered and left.

The fox-trap which I have figured here we found about two miles below our camp. It is really a form of dead-

fall trap, and was baited with seal-fat; but though foxes had been about it I could find no evidence that any had been lately caught. The trap was not set, and I think the bait had remained there from the winter catch. For there is no value in the fur of a summer fox.

Had we not been better informed, the mirage to-day might well have deluded us into the expectation that our friends had come. For, as we walked home, there appeared on the cliffs in front of us a high flag-staff with a brave flag floating atop. But it dwindled as we came up, as well we knew it must. For it was nothing finer in reality than a little three-foot stick with which, when I had tied my handkerchief to it, I had marked the site of a nest.

When I had pressed the flowers this evening I sat for very long watching the red-throated divers, of which many drop into the creek below the tent when the tide is at half-ebb.

They were only taking one kind of fish, the Arctic flounder, which the Russians call 'kambola.' These they caught very rapidly and never missed. The fish are not swallowed until they have been worried well. I never saw one diver try to rob another as ducks will; there was enough for all, and mutual respect.

I, too, had designs upon these flat fish, none of which were to be realised. It was not a little trying to see these birds gobbling down with so much satisfaction a kind of food which would have exactly suited us. But tired as we had become of eternal long-tailed duck, I could contrive no way for varying it. Had we but been possessed of our dredge, or even fish-hooks, I doubt not we should soon have filled the pot. But these appliances were upon the yacht; and though I schemed a method for taking flounders with a bent pin and a bit of string, I found that, without a boat, or at least without a fishing-rod to reach over to the deeper parts, the deep black mud of the creek was too great an obstacle. Had we really been pushed to a necessity I of course could have ventured on mud-pattens made of drift-wood, but we were not yet in that case. Another time I should, I think, take Indian snow-shoes, the best contrivance I suppose for work of this kind. They would be most useful as a support when walking in the lakes. In these, though there is no danger of disappearing altogether, your feet often sink so deeply as to tire you very much.

July 14th.—Ice unchanged.

Now we had one of those rare lovely days which redeem these regions from the charge of monotony.

The south-west wind held all day, and when I went out to the long creek, before five in the morning, the tent thermometer showed 60° even at that early hour. At 3 P.M. it registered the same, for the wind had grown a little, and at 8.30 P.M. it had dropped to 41°.

I did nothing at the creek, and was back pretty hungry by eight o'clock.

In the afternoon I struck inland and found the lakes you see on the map. Here I secured four long-tailed ducks at one shot, and got horridly wet through lying in the reeds while I waited for them to come within distance. I found that by moving my cap about they would come through curiosity; Sailor was also a good draw sometimes, the long-tailed ducks following him round, as the ducks follow the dog of the decoy ponds.

Musquitoes, which regularly appear when the heat reaches 60°, were a nuisance in the swamps.

Red-throated divers were mewing and braying like

idiots; most birds behave foolishly when courtship is on, but these were passing words.

I only found a single willow-grouse. There could be no doubt that we were on a very poor ground for these birds.

The accompanying sketch (p. 206) I made at midnight, while Hyland went up the long creek. Here he thought he saw some geese with young, but could not come within shot of them.

On comparing notes this evening we both found that our hands, which had been so much swollen upon our walk, were now shrunk to that extent. The skin seemed almost dead, and was so shrivelled and loose that I can compare its appearance to nothing better than a baked apple. Also, we had lost all feeling in our finger-tips, except a numbed sensation like 'pins and needles.' This curious state I could only account for on the supposition that it was due to an abnormal condition of the nerve terminals from excessive exertions.

Sunday, July 15th.—The ice all day boomed like cannons, and was moving up, so that we began to have good hope that the *Saxon* would be able to come round all right.

The sun was very hot at mid-day, and the thermometer rose to 70° in the tent.

All day long great flocks of king eider, long-tailed ducks, and dunlins were crossing the sands. For the first time we had musquitoes in the tent.

SCHAROK AT MIDNIGHT

The glaucous gulls had left the huts; they only came now at rare intervals.

On this day something happened to my inside. I had gone up the long creek with old Sailor, and lay down for some time watching some grey plover. When I rose some pinch took hold of me which pretty well doubled me up, so that I could but just succeed in crawling home.

It seemed to me that we wanted vegetables. So we opened a jam-pot which we had been treasuring against a famine time. The jam said it was strawberry, though it seemed like glue. But we ate it, and then I turned in at 11.30 P.M., while Hyland went up to the big lake to try for a duck.

He had better luck than that, returning with a couple of white-fronted geese, and a little living young one. This youngster, he said, dived splendidly, but none the less Sailor caught it alive, going clean under after it.

Our gosling had a way of standing straight up on end, but with his breast puffed out, very consequential, exactly recalling a city waiter. So 'The waiter' he was named.

He really was a dear little bird. Not in the least bit shy of us, he inspected every corner of the tent, and at last went to sleep for several hours under my chin.

July 16*th*.—Ice still and apparently unchanged. Musquitoes very bad again, though the day was not nearly so hot.

The waiter began the day's undertakings, and we had to follow. He was out from under my chin at an early hour and away exploring. He walked up and down me for quite a while, but left that presently by the bridge—my face and head—and so reached the big tin case, which was a finer promenade. Here he marched up and down for a little while like a captain on the quarter-deck, cocking his head on one side in a very comical way. While considering how he could cross to Hyland he ventured too near the edge of the cliff and dropped right into a ravine between the tin case and the biscuit-box. I drew him out of that after many attempts, and started him again. This time he attacked Hyland, as ravens attack a dead body, going for the eyes. But Hyland was only sleeping: till this woke him up. And when we rose the waiter became a serious question, for if there was a place where it seemed you could tread with safety he was sure to sit down there. So then I caught him and penned him up. Many times I penned him and in many ways, but always he came out at a run.

Then Hyland said he would build a proper pen. And he made one that even defeated the waiter. For fixing a circle of reindeer horns, he covered them with slabs of moss which curled inwards at the top.

While this was preparing the waiter came out walking with me, and betrayed a great liking for the green heads of cotton grass, which he nipped of neatly just below the bud. Equisetum also he went for greatly. He disliked to be left alone, running quickly after us like a little dog.

We picked this day a bit of wild camomile, just coming into flower; Hyland found it growing on the roof of Uano's hut; for we climb up there when the weather is clear to see how the ice is looking in the gulf.

Black-throated divers were taking to-day a large and new kind of fish. I could see them with the glass, but not what kind they were.

Hyland, who was out in the evening, brought back two interesting birds. The first, a curlew sandpiper, he shot among a lot of dunlins on the mud.

This bird you will like to know something about.

It is small, eight and a half inches long, and its back is curved rather like that of the common curlew, whence the name. These waders pass through our country in spring on their way to nest, and return with their young in August and September, when they remain on our mud-flats for some little time.

Yesterday, July 15th, twenty years ago, Mr. Seebohm shot one out of a flock on the Petchora, and Dr. Middendorf, in June, shot a bird on the Taimyr Peninsula, which would very soon have laid, for in the oviduct was an egg partly covered with shell. That is as near as we have got at present to a proper view of the egg. Where does this bird nest? No one can tell with certainty.

We were not to find it nesting. This was the only

individual we saw, till we got one later on in winter plumage.

The second bird was alive. It was a young turnstone, some seven days old. We put it in with the waiter, who opened his mouth at it and made faces, but would not be friends.

July 17th.—The ice unchanged. A hateful day. The thermometer at 40°, with a driving fog and north wind. This fog, of which I have spoken so often, is really a 'Scotch mist,' which falls on the tent like rain.

If you leave things out in the sun you must not go to sleep. The wind here goes round in a minute and all is up. Now it is sunny and drying, with a warm wind from the south; suddenly the wind chops round to the north, and at once a fog, cold and wetting, sets in.

We had a seal or a young walrus in the harbour to-day. It showed several times.

But I did not get about much to see things, for I had *la grippe*, or something akin, which pinched. I lay up chiefly and blew eggs. Hyland, who had never before known me unwell for a moment, began to get alarmed. He was exceedingly kind, and 'boneing' some turnstones and grey plovers, cooked them beautifully for me.

At half-past five in the evening Hyland was plucking a long-tailed duck, the waiter was making bluffs at the turnstone in the pen, and I was blowing eggs, when all of a sudden up comes Verrmyah Uano, with a couple of teams and sleighs.

He was very cheery, and glad to get some tea. He said he wanted me for the goosing.

Now see what paradoxical things we are. A moment before I had been blowing eggs out of pure bravado; indeed I felt more dead than alive, for I had a real chill on my inside, and a sort of unkind cramps. But now nothing would do but I must go to the goosing, come what would. I said to myself, 'I will be a Samoyed.' So, beyond a bit of macintosh to lie on, a pot of Liebig and a biscuit, I made no provision for the stay, which I promised Hyland was to last but a few days.

I would have liked to have made it longer, but I knew that my companion would not be happy for long alone, as he had not those resources which keep one from the dulness of the solitary life. 'Also,' I reflected, 'if I am away, and he gets upon him the ague or the giddiness, how will he fare?'

So I left.

CHAPTER II

THE GOOSING

We now had thirty versts to travel if we were to make our old Pesanka camp.

All went well till we reached the Baroshika, and here we came to an enforced delay. For the river had risen badly and was very deep. We tied the reindeer's heads to the sleighs and tried the river with the 'toorr.' It was everywhere unpromising. At last we could delay no longer, but determined to cross at our old ford where the water would not be higher above the normal than two feet.

So we slung our belongings round our necks, and standing on the sleighs, prepared to cross. I doubt not this sounds simple enough. It was not. It was easier for the Samoyed in his seal-skin boots. But to stand on a small, slippery, moving sleigh in nailed boots, with water rushing round your feet, is hard. The drop from the bank into the water is the first trial, and the bumpings, risings and sinkings of the sleigh over the sandy shallows are worse. However, we did get over with water up to our knees, and we crossed without other mishap than a broken sleigh. For somehow my

A HALT

Face page 212

bar snapped, but Verrmyah, splitting a thong with his axe and fitting it in, contrived to jam the tenon up.

We had much the same difficulty with the Peinmur river, but rose at last the south side of the plateau, on the northern edge of which we had lain in Pesanka camp. And here we came almost at once upon a very different and unlooked-for scene. It might almost have been Derby Day in a corner of Epsom Downs.

I looked in vain for Uano's choom. Instead of it I found a rough semicircle of gaily-coloured gypsy tents, blue, red and double-coloured, and not a single Samoyed I knew.

Men, women, and children were scattered all about, busy as bees as usual; but they were all new faces.

'Where is my choom?' I asked Verrmyah, who had told me by the way that they had a beautiful place for me.

'There is no choom; it is a tent;' returned Verrmyah, and roared with laughter when I looked surprised.

With that he leads me to a little tent, and pointing at me, simply says, 'Pud-drr'—'yours'; or more correctly 'thou.'

(Notice the word Pud-drr, literally 'he,' is used also for the *second* personal pronoun, but then is accompanied by a point at the object. In this way the Samoyeds, like most civilised peoples, avoid the use of the direct 'thou.')

So I lifted up the flap and looked inside. It was not encouraging.

You see it was close on midnight, and Uano and old Ustynia had gone to bed. Their tent would just hold three sleepers, cheek by jowl. And, as my entertainers, they had laid a special white reindeer skin just between them, which was to be my bed.

'Sdorova?'—'How d'ye do?'—they said without rising, and then tapped the skin between them and said, 'Here your bed.'

Well, I didn't want to hurt these poor people's feelings, but I felt I wasn't up to studying Samoyed life quite so closely as all that. I laughed, and said it was too hot to-night, I would sleep outside; and, under cover of this, retreated.

I found an old boat, which, turned on its side, would do very well, and under this I spread my macintosh, and lay down to sleep.

Now see how kind these people are. Rain-showers (and nasty ones) had come on, and yet I saw Katrina and little Zornka busy about the fire in the open, and Verrmyah plucking a goose. But I didn't take much notice beyond wondering that they should trouble to do that in the rain.

But after a bit little Wanka comes up to the boat with a wooden bowl of smoking hot goose, saying 'pillet' (*i.e.* plate), an English word he was very proud of. But I was afraid to eat goose-stew at that moment, so then he brought me 'yah'; and I was afraid to drink the 'yah.' For I knew that if I was to go goose-hunting to-morrow

I should have to look after the cramps. They were very much disappointed at my refusal of this, which it seemed they had made especially for me. I told them I was not well, and then I mixed some Liebig and drank this most excellent concoction, following it up with water nearly boiling, which made me feel better at once. The goose they insisted I should keep to eat in the morning.

Also I gave Mekolka Liebig. Uano said Mekolka was ill—ill in his chest—and coughed, with symptoms that pointed, I thought, to incipient consumption. And poor Mekolka, saying nothing, looked so wistfully at the Liebig, which he doubtless imagined was most wonderful medicine, that I could not choose but give him some, for which he was most grateful. I mention this because one does not always look for a capacity for fine and enduring qualities such as gratitude in a poor half-savage man. But from that moment Mekolka was attached to me in a manner most affecting. He began to show it at once. For he fumbled round till he had found an old sail, which he spread over the boat to keep off the wind and rain.

The little boat under which I lay was just the ordinary 'arnoh,' which these people make from drift-wood. I think I have described it before.

For a long time I could not get to sleep because of the dogs. They were not *our* dogs—which were always tied up at night—but tramps, belonging to the other

Samoyeds. For we have two stranger families with us. Of whom more to-morrow.

A wandering dog kept pushing in his nose under the sail against my head, on the look-out, no doubt, for goose. For a long time I took no notice. You learn to be very careful with these dogs. One does not matter much, but you may very easily bring the whole pack about your ears. At last I thought it had gone far enough, and the next time the canvas begin to move I knelt up very quietly, doubled my fist, and caught the prowler a crack

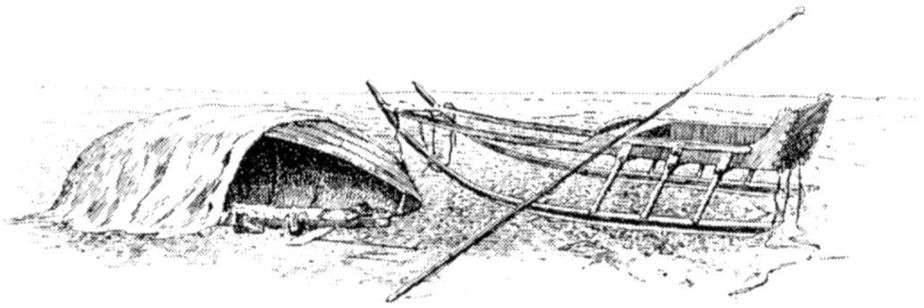

MY ADLIURS, MY TOOR, AND MY RESIDENCE

on his nose so unexpectedly and with such effect that he fled half howling and half growling. A Samoyed does not hit his dog, he only adjures him; so no wonder this dog was scared.

With that the whole pack—and I reckon we had thirty dogs—broke into a scream like maniacs, and the women rushed out to see what was up; so that, what with the noise of the dogs and the women, and the men who yelled from the 'yierserk' [tents] at the women and the dogs, for a short time we had quite a Wagnerian symphony. But gradually all grew still. Once only after that I heard a dog creep up with a half-scared, snorting sniff; but what with the strange smell and the sudden hidden blow, he probably concluded it was the devil, and troubled me no more.

All was still, or would have been but for the reiterated views of Mrs. Uano, who chid her lord for snoring. And indeed it came as a trumpet-note across the twenty yards or so that lay between me and them. But at last I slept.

July 18th.—Instead of the drear north wind of yesterday we had a keen wind from the east, but it was distinctly warmer. The rain had tailed off into a dripping fog, and I woke up very glad of the boat, for outside all was soaking.

A Samoyed keeps much the same hours as the editor of a daily paper. When I went out at 10.30 A.M. not a single soul was moving.

The cramps were gone, and I never felt fitter in my life. I ate my goose, and then, when the fire was lit, followed it up with hot water.

For the fire was lit at last, and all the camp sprang into activity.

I received various calls, and began to find out who was who.

Old Yelisei came—the father of the island, he said he was, and seventy years old, with a fifty years' residence on Kolguev, or on 'Awh.'

I do not know how to write the native name for

Kolguev; it is somewhere between 'oh' and 'or,' pronounced very shortly as an exclamation.

And Yelisei's wife was there. She was stout, and wore quite a handsome dress of reindeer and dog-fur, with splashes of red, yellow, and black. She was a born stage-duchess. With them came their son Serieh, a nice, cheery, good-looking man.

The third 'choom' (understanding the word in the sense of the family), numbered more. Its head was Marrk. Marrk was a kind old fellow, but I have an idea his head was going. He had curious lapses into abstraction, and did little work. With him was a charming young fellow, Kallina by name, whom I took for his son; but you shall hear more about him later on. Two daughters also came—good-looking girls, with a wonderful collection of beads and brass about their hair; they seemed of the same age as the elder girls of our choom.

In addition to these we had certain unattached items—Onaska the prophet, for example, and another man.

I leave you to imagine the curiosity I aroused. The first Englishman they had ever seen or heard of, and one who had come so strangely there. For you may believe me when I say that 'Englishman' conveyed nothing to them. Hyland, indeed, they knew as 'Hylum,' and me as 'Ahnglia.' I had told them our country was Ahnglia, and they thought it was my name.

I again fed Mekolka on Liebig, in return for which he stopped some of the chinks in my boat. My friends were filled with wonder that I should choose to lie under a boat. As a matter of fact it was far better than their wretched 'yierserk,' which let in all the rain.

At about four o'clock we set off for the goosing, and had much fun by the way, for what with my prismatic compass and route-surveying, my flower-picking and questions about birds, I was a mystery and the subject of constant wonder and amusement.

The new-comers were rather shy about my camera, dodging behind each other in dread of snap-shots. But all *our* old friends jeered at them, for they had long since got over their mistrust. To them it was no spell from Nûm, nor an application of the evil eye, but some silly thing which amused me and did not hurt them. I had told them they were sun-pictures; so whenever they saw the camera in my hand they would look from me to the sun and from him to me, expecting I don't know what. I was always so sorry I could not show them a picture; but we had no developer and no dark room, so I gave them sketches instead.

In this way we reached the coast. I should like to try and give you a clear idea of the goosing or goose-taking, because I do not know that it has ever been told before.

The scene: A low-lying stretch of land, half peat, half grass and marsh [running north ten miles to the Pesanka hills, and south five miles to Barokoska Nos], and an island on the tidal flats some four acres in extent.

Round this, now that the tide was out, sand or shallow water, which deepened to a wide creek against the island's southern bank. Beyond this again sand or mud in ridges, and creek after creek. Further yet, perhaps three miles to seaward, the long line of the outer sand-bank with its piled-up ice—and then the sea.

So bad was the day that only now and then as the mist lifted could you see the farthest ridges and the higher banks of mud. But when the banks appeared they were crested with a *cheveux de frise*, which you knew were serried lines of geese.

Seven boats under the command of the younger men were soon slipping down the creeks; for they were to get behind the geese.

Then the reindeer teams were driven out, three on one side and four on the other, remaining as near as possible equidistant, to prevent escape by the flanks.

All were now away except Uano, his wife, Katrina, two or three small girls, the little boy Wanka, and myself. Katrina nursed her baby.

Before half an hour was gone by the geese began to rise. We could see them through the rain getting up in hundreds off the sands.

Uano sat by my side and muttered. 'Bad big geese,' he said, 'big geese no good. Little geese good. Big geese fly,' and he flapped his arms like flying.

Yes, the grey geese were evidently not yet moulting.

In his hands Uano held his 'parlka'; so they called

it, using the Russian word for a stick. But it was a very formidable stick. About two feet long, it was shaped like a policeman's truncheon, only nails or bits of iron were driven through it all round. Uano soon let me see its object. For an old grey goose, who had somehow reached our island unperceived, now appeared in front, with head straight out and carried low down, running wonderfully fast among the grasses. Presently it squatted; and then the Samoyed crept up on tiptoe. I followed. The goose never moved till the man was ten yards off—I saw it squatting there with its head on the ground and its neck stretched out to its full extent—and then it rose and ran. At the same time Uano hurled the 'parlka,' and with such effect that the bird was knocked out of time altogether. It was all but dead when we picked it up—a male bean goose, which had just lost its long wing-feathers.

It is well to explain that all the grey geese belonged to one of two species. Either they were bean—the commoner goose on Kolguev—or else white-fronted geese. The pink-footed goose is not found there.

Away behind us on the island was the trap. I must describe this particularly. At the water's edge, thirty yards apart, two poles were fixed, to which a net was fastened. The net was then carried inland, the two walls converging until, at a point some forty yards from the entrance, they were not more than five yards apart. From this point they bellied out and formed a circular *cul de sac*. The netting was about four feet in height, of some three-inch mesh, and round the *cul de sac* was double. The uprights which carried it were strengthened by spurs.

A net trap of this kind the Samoyeds call 'Po-ùm-ga.'

Long before we could see the boats, for the mist had thickened, we could hear shouting and the cries of the geese. But after a bit first one and then another boat came into view. On the men came, but very slowly; now pulling across a creek, now pushing the 'arnoh' over a bit of mud or hauling it over a sand-ridge, sometimes leaving it altogether and running off to head the geese. So slowly they came zig-zagging along.

By this time we could see geese by thousands through the mist. I could even distinguish the short trumpet-note of the brent among the general babel. It was indeed a babel. How to convey to you any idea of it I do not know. If you can imagine many hundred farm-yard geese, and many thousand cornets all sounding together and crowded on by a handful of screaming wild men—if you can imagine this, then you are not far off the mark.

On they came. Now I could see big grey geese running, heads up and wings outstretched like any farm-yard geese, breaking away to the right and left. And now I saw that they had small young ones with them.

Nearer they came and nearer, the middle a dense

solid mass of geese, the sides a constant stream of parties, large or small, running away like lamplighters for all that the sleighs might do to stop them.

The very earth seemed geese, and for that matter the sky too. For there never was an interval when geese were not rising, and instead of going right away at once, as one would have looked for geese to do, they hung about the spot, circling round and rising higher and higher till they lost themselves in the mist. I could never have believed it possible that so many geese could be had on one small island.

And now I became aware that there were geese with us. Invisibly somehow they had arrived; but there they were, slipping along in the old way, heads out and low, and squatting in the stuff.

Exactly at nine o'clock—five hours from the beginning—the advance guard of the swimming geese came round the corner of the creek. It was one solid phalanx of brent. They seemed to be by far the fastest swimmers. For behind them at a considerable distance followed a smaller lot of grey geese, some swimming, some running along the edge.

Then with one accord nearly all these grey geese rose—five hundred perhaps there were.

Uano and I were lying low in the grass about fifty yards from the water-side; but for all that I contrived by looking over a little mound to keep an eye on the scene.

For some little while the geese delayed as though

they felt they were getting too much inland, or suspected a trap in front. Then the boats came up from behind and the geese crowded on.

They didn't like going. Sometimes the leading geese would stop and wheel about, heading right into the mass. But the boats came on.

No sooner had the boats passed our position than we rose, and Uano set to work at once to polish off the few that had hid themselves about us. He slew some with a toss of the 'parlka'; some that lay till almost trodden on he knocked on the head as they sat.

He couldn't make me out. 'Quick,' he cried, 'kill them quick. Uano old; you young, you. Good goose, goose good, very good'—but I could not harden my heart to kill wild geese like that. So he shouted for Ustynia, his wife. And Mrs. Uano, ordinarily so composed and kindly, came flying across to us like a mad thing. The way that woman went on! She hunted round like a hound, she hurled her 'parlka,' she dropped full length upon the squatting goose, and rose triumphant and screaming with laughter with the goose in her embrace.

But the boats came steadily on. Every moment I looked to see the brent escape by diving, or expected some to rise, for it was plain enough that many were full-winged. Neither of these things they did; only like a pack of idiots they 'wanked' and swam along.

But the grey geese dived. The bean and the whitefronts behaved exactly alike. First they laid out their long necks flat on the water as their fellows did on land. Then, as the boats came nearer, they sank their bodies till the water was almost over their backs. It was wonderfully difficult to see them then—they looked like bits of stick.

When a boat approached a bird it would just sink its head and shoot forward under the water. They never went down like diving ducks.

And now the body of brent was exactly opposite the entrance to the nets, and about them in a half circle were the boats. Round and round they swam, but refused to leave the water. The boats did not dare close in for fear the geese should break. It was a ticklish moment—the geese would not make the land.

At last a single old goose—a bean he was—stepped out and ran up the bank. He was quickly followed by one or two more, and then by the first of the brent.

And now that they had started they went quickly enough, scrambling after one another and heading into the net. Over the green they ran like a flock of domestic geese. Sometimes they aimed for right or left, but then the children showed themselves and the geese were turned.

The last bird was in, and then we closed the rear. Not a brent had flown, not a brent had dived, not one escaped. Of all that army every bird was in the net—a dense, black, moving mass.

Then the way to the net was closed by lifting the pegs and moving the net inwards. The women and children kept guard outside, while the men entered the circular apex and began the killing. In the meantime a kind of hurdle had been placed against this narrow entry to keep back the crowd of birds.

The killing was a long business, though the actual details were done rapidly enough. Each goose was picked up by the head and swung quickly round, by which method the neck was broken very soon. Then it was thrown out of the circle on to the ground outside.

Meanwhile the geese of the outer ring were running round and round, every now and then rushing at the net. Occasionally one managed to get out, and then the excitement was great. Little Wanka had quite a chase on his own account among screams of laughter from all the Samoyeds, and after many tumbles and sprawls brought back his bird triumphant.

It was really a sickening performance, though one realised of course its extreme importance to these poor people in view of winter food.

Now and then a white bird caught my attention, and when the numbers were reduced I managed to secure it —a brent, almost entirely white. Unfortunately its primaries were moulted out.

And so the grim game went on. Each time the inner lot were despatched the door was opened and a fresh batch admitted. At last there were not more than thirty

DRIVING THE GEESE

Face page 226

birds or so left outside, and then the men stopped killing and the children were turned into the big ring, while the elders stood outside, and each backed their own offspring: who could catch the most—a kind of game—one choom against the other. So the children screamed and snatched and sprawled, while the men shouted and encouraged, and the women laughed fit to split their sides. Poor geese! they made a good show for it, but they were all caught at last.

I cannot say I liked all this. On the contrary, I felt very unhappy. But there it was.

Then the counting began. We had taken in the net alone,

	3300 brents,
	13 bean,
	12 white-fronted.
Total,	3325 geese.

Onaska kept the record, as the counting went on, with notches on a stick.

Then the geese were again subdivided and apportioned out in heaps, so many to a choom. One point of much interest for me was the fact that both our varieties of brent, the light- and the dark-breasted, were represented here.

July 19th.—It was so late by the time all this was finished, being indeed somewhere about two o'clock in the morning, that it was decided to leave all stowing for a second visit.

But before we returned there was a little ceremony whose quaintness struck me much. Five men came up to me one behind the other, with old Yelisei at their head. In each hand they carried four brent geese. Old Yelisei laid four down on the right of me, four on the left, saying solemnly in Russian this, 'Yelisei's gift to Ahnglia. Little goose, very good, very good.'

So I shook him by the hand. 'Thank you, Yelisei—very kind,' I said.

And the next man came with 'Yugora's gift to Ahnglia. Little goose, very good, very good,' and Yugora retired. After him came the rest in order, each with the same formula as he delivered over his load. It was quaint to see and quaint to hear. No doubt they had agreed beforehand what they would say—and it was quite a big speech for them.

So I now had forty geese. But I was struck by the fact that old Marrk had not come.

Then we left and had great fun going home. For the women said they would be home before us, so that the fire should be ready. With that they left and started racing, the children urging on their four reindeer to their utmost speed; and as they were the lightest they distanced the women altogether as far as we could see.

When they were out of sight Mekolka, son of Uano, beckoned to me. 'Come,' he said, 'Mekolka's reindeer very quick.' So we jumped on our sleighs and went off first at a trot. No sooner, however, had we risen the hill than Mekolka turned down into a river-bottom where the snow was lying in broad belts. Here we put the reindeer to the gallop, and sometimes on the snow, sometimes on the river shallows, kept up that pace the whole way home. All this time we were hidden from the other travellers, who were moving over the higher peaty land.

We reached the camp, among a chorus of yelping from the dogs, nine minutes before the women and children had arrived. When they reached the brow of the hill they found to their astonishment the smoke of the fire curling up. This was a very good joke. They laughed and talked about it for some days after that.

One curious result of the want of the definite article in the Samoyed is that articles in daily use seem to become personal. Just as a yachtsman talks of his boat—*Saxon*, *Valkyrie*, *Arrow*—as though she were a human being, so is it in Samoyed with everything. Thus—

the fire—too
the bow—ürn
the boat—arnoh
the big pot—yud.

'*Too* wants wood,' '*Yud* is empty,' they say. So that after a bit one comes to look on 'too,' 'yud,' and the others as persons rather than things.

I was now to all intents and purposes a Samoyed; eating goose with my fingers, drinking 'yah' from the bowl. They had indeed the advantage of me in eating, for they could use their knives with the upward stroke, and I did not dare for fear of losing my nose.

Even the dogs accepted me, and that I thought the ultimate test. They were quite quiet now when I walked about the camp. Probably I had begun to smell more 'Samoyedy'—homelier by now.

We had a good deal of fun before we turned in. But this will be a good place for saying something about the baby—'adski,' as they always called it.

Baby—it was Katrina's—lived in the normal way in a drift-wood case like a lidless box with rounded ends.

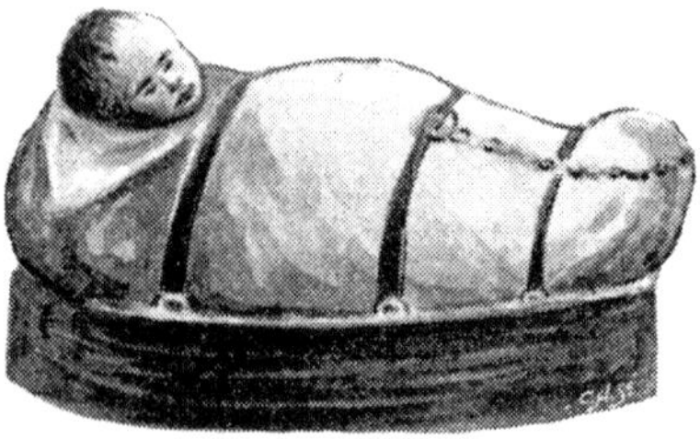

Into this was first put a layer of reindeer skin; then came a layer of dried sphagnum or water-moss, and then came baby, with its arms by its side. After baby was another layer of reindeer skin, then a quilt or sheet of coloured cotton. Then round this were passed brass chains and straps to tie the whole together; and there you had a compound organism the nucleus of which was baby.

So, like an Indian's papoose, the whole structure was carried about, and when the child was jumped about or nursed the whole apparatus shared in the movement.

Baby, of course, could not move while in the pupa stage, so to say, and was only liberated at rare intervals. Then it was extracted—a curious chocolate-coloured little creature—and danced about in front of the fire without a stitch of clothing on to keep it from the draughts.

The girls had no greater delight than in nursing and playing with the child, and little Wanka loved it too. He used to kiss it in the Samoyed way—that is, by a pressure of the lips without the sound of kissing.

The Yeleseis and Marrks went soon to bed when we got back that morning, but we sat up long, talking in our choom. They wanted for the hundredth time to find out all about the land I lived in.

It was farther than Norway I told them. 'Yes, yes,' said Uano, they knew where Norway was; they had kept and fed some Norwegian sailors who were wrecked on Kolguev some years ago.

'Yes, yes,' adds Uano, 'and Norwegians say, "Uano very good; very, very good. Russian not good. Uano *good.*"'

Then they said it must be the land where they killed poor Samoyeds with knives.

'Oh no, indeed,' I tell them; 'that is not true. Will they—will Verrmyah, Katrina, Ustynia and Uano come back with me; I will take them?'

'No, no, Ahnglia stop here. Ahnglia always stop here, and Hylum stop.'

With such simple talk we beguiled the time. And then I made Uano sing a hunting-song or saga. I had heard it often before. It begins: 'Mein Neydòo adliurs hunyan?'—literally, 'My son (his) quick-sleigh where (is it)?'

I made them explain it in Russian as well as they could, and got at the sense very fairly.[1] Only they would all talk at once, and squabbled about the proper equivalents for their words.

I think it might be rendered like this :—

MEIN NEYDÒO

Where is the boy's small sleigh? He is off—
Gallop and speed, gallop and speed—
Gone with the wind by the hunting-pass,
Gone with the wind to the icy sea.
He that can bend his father's bow
Leaves to the women the reindeer call—
Killoo, killoo, killoo, killoo!
Leaves to the women the fire and pot.
His are the best of the four-year bucks,
His is the fastest team of five.
Soon he will come with hunter's gains,
The water-bear and the walrus dead,
For the winter bed and the lamp of oil.

Show me the peer of *Mein Neydòo.*
See, he can drive in the path of the night
Straight to the choom from the island's rim,
Straight in the night with the team of five.[2]
All that his father knew he knows;
Knows where the swan *horeh* will nest,
And where *wurrah*, the small black goose.
Yes, he can set the trap with skill,
Fashion the arrow, frame the sleigh,
Can *Mein Neydòo.* Is he wise?
He is wise.

[1] They are capital actors, and emphasize each song with such expressive pantomime that after hearing it two or three times it is no more difficult to follow than *L'Enfant prodigue.* In the last act the reciter throws himself down flat on his back.

[2] The Samoyeds are very proud of their power of taking a bee-line for home in the dark.

But he can toss the drift-wood log,
Throw the dizha, take the deer—
Take and hold the largest deer—
Can *Mein Neydòo*. Is he strong?
He is strong.

Where is the wife for *Mein Neydòo*?
Nya, the maid, for the man *Nyanìtz*?
She must be all that his mother was,
Deft with the needle, skilled in *yud*;
Too—the fire, *yud*—the pot,
Yi—the water—these are hers.
Obah—the glove, can she make it well?
Nier-pi—the seal, can she dress the hide?

Yes, there is one; she is come—she is here.

Lay me down on the tundra now—
Old and worn, old and worn—
Leave me alone on the moss to die.
The broken sleigh, and the spoon I used,
Carry them up to the Holy Hill,
And speak as I speak: 'Oh Nûm, Great Nûm,
He lies on the moss quite straight, quite still,
Who once was wise and strong.
Many a time has he vexed thee, Nûm—
Yet now, Great Nûm, forbear—let him rest.
See, it is true, the spoon, the sleigh!
They were his—they are thine—He is dead.'

This would be sufficiently like it. And the tune? That was the worst part of it. Only after hearing very many of these songs could I at last come at any method in the tunes. In general they were not unlike bagpipes, returning often to a fundamental long-drawn note.

They loved singing these songs; you only had to say 'Huntz,' *i.e.* singing, to set them off at score.

And then they would dance, singing an accompaniment; such a stupid old dance, round and round, with great lifting of the arms.

They could not whistle,[1] and envied us very much this accomplishment. The girls used to practise it by themselves, till Anka at last became a very tolerable whistler, and would whistle incessantly as she drove her sleigh. Then they would know whether in our country we had chooms or huts. And when I told them no chooms, only huts, they thought that was a very bad plan. For they remembered how we had explained that we lay down without any cover on our way from the Gusina. And they said if we had no movable house at home, then, when we could not get home by night, we must sleep on the ground, which was not good.

We got to bed at 4.30 A.M. A dripping fog had come on, but my quarters were fairly dry, for most of the wet had been caught by the sail. The camp did not wake till eleven o'clock. I was out a little earlier and prowling round. Particularly I was interested in three little blue foxes in the Yeliseis' camp.

It was the hottest day we had experienced. Even the dogs were beat by it. They puffed and panted, and, sprawling all about, refused to budge.

Poor old Pasco nearly succumbed under the combined attack of her six puppies. Here is her picture taken on the spot. Pasco was a very pretty bitch, with long brown fur and a lovely brush, but her puppies were of

[1] Yet they had a word for whistling. They called it 'Shu-ga-ring-on.'

all colours. They were beaten at last, and rolled over, too fat to move. While we were in 'goose camp,' and after, they fixed up a little choom about four feet high, and here Pasco and her puppies lived, and at night this shelter was also shared by the two younger girls. How they all packed in was a mystery to me.

Pasco loved those pups for all they were so hideous—and they were. Whenever I passed too close she had a way of running straight at me quite silently, as if she

were going to bite. But she never actually bit me, though she came very near it once or twice.

We had a little brush to-day, which made me sorry. I had put my cap on the top of the boat with other things to dry in the sun. On coming out the cap was gone, and there was Uano strutting about roaring with laughter, with my cap on his head. Now take a Samoyed as you may, you would draw the line at lending him your clothes. The old idiot began dancing round, saying, 'Uano Ahnglia to-day.' But he suddenly dropped that. He had never seen me look angry before, and I think it scared him; though I had only said quite quietly, 'Put it down.' He obeyed like a lamb, putting it on the ground exactly where I pointed my finger. Poor old chap! he looked at me deprecatingly, as your retriever will sometimes look. 'Uano good, very good,' he said. But I felt I must give him a lesson somehow. So I said, 'Bring some water.' He fetched it in a wooden bowl, and then I made him pour it into the cap. This done I swished it round, emptied it, and said, 'Now put it on the boat again.' He did it all so humbly and obediently that I fairly wanted to smack him on the back with 'All right, old fellow, no bones broken over that.' But I felt that this impression might be of lasting use in case we should ever have to leave our things unguarded. So he crept away to his choom.

I was sorry, and instinctively knew I had lost ground. If they once began to fear me they would retire into themselves with all the old reserve they showed at first, and then good-bye to my Samoyed studies. This came home to me forcibly at breakfast-time. Instead of little Wanka coming running down with 'Ortow Ahnglia!' at the top of his voice, as he usually did, there comes old Uano very solemnly, with a wooden bowl of goose and salt and bread. This he offers to me as though I were the great god Nûm himself. I took it for this once.

The heat beat us all. The men and women put off their skins for the first time, and came out in red shirts with the knife-belt round the waist, and the shirt tails outside in Russian fashion. Young Yelisei had a green

THROWING THE DI-ZHA

Face page 237

waistcoat (of all things) over his shirt, and with his long black hair and thin moustache looked just like a super in Hamlet.

We entered to-day on a new phase of reindeer life. For the first time the fly appeared (*Hypoderma tarandi*), known to the Samoyeds as 'Pì-liur,' and to the Russians as 'Orwot.' The first I noticed was probing the cracks in my boat with its ovipositor; no doubt the grease

1. REINDEER FLY (*Hypoderma tarandi*)

2. SAME, WITH OVIPOSITOR EXTENDED

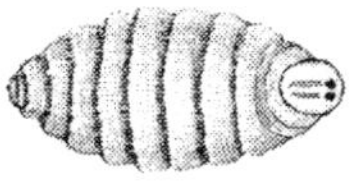

3. LARVA OF SAME

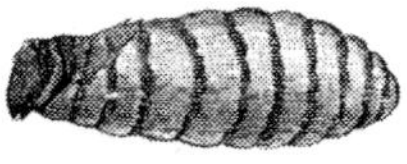

4. ADVANCED LARVA OF SAME

would make a fair nursery. This fly, which has a general resemblance to a humble-bee, lays its eggs on the back of the deer, where it is out of reach, and these eggs develop into larvæ ('Oograh' in Russian, 'Sivia' in Samoyed), which, burrowing into the creature, worry it terribly. The deer have therefore to be often caught and examined; and this catching is done with the 'di-zha.'

The di-zha is simply a lasso or lariat of plaited hide, the noose of which runs through a block of reindeer horn. I practised throwing it at a stump. The Samoyeds throw it underhand, and not so well as the cowboys with their overhand throw. The deer are brought up by the dogs and the adliurs, and are then divided into two lots. The cows and calves are constantly bolting from one side to the other as the men creep after them, and on a favourable chance the di-zha is thrown. A calf or young deer makes a fine struggle, and takes a lot of holding, but very often the cows, from having been often caught before, come in to hand without any or with little resistance.

The poor deer tormented by the fly rushed all over the place, scuts up, snorting and grunting.

The musquitoes were bad to-day, and gnats were in clouds. I think the Samoyed name for the musquito is so good—'nyanink', they call it, from the noise and its stab. 'Nya' is its singing, and then 'nink,' and in goes its horrid little stab.

However, at 11.30 we left again for goose creek.

First the sleighs were loaded on the island, and then the geese were brought across to the mainland and cached.

The turf cut round with the axe, where the cloudberry grew thickest, was torn up with the hands; then the geese were stood on their tails with the heads tucked in, till the girls had made a circular group some three or four yards across. Then the turfs were rolled back on them a double layer, and the packing was complete.

'Dorndaftsa come hungry now, and dorndaftsa go

THE WOMEN CACHING THE BRENT GEESE

Face page 238

away again. No get in; no good. Yes, yes,' says Ustynia, with a chuckle. 'Dorndaftsa,' by the way, is our old friend the blue-bottle.

I can recommend this plan for any who would keep game in this country.

We had passed within a yard of a king-eider duck's nest a dozen times, I should say, without seeing it, when she suddenly rose straight off her eggs and away to the creek. The extraordinary point was this, the eggs—there were five, all incubated—were completely covered up; and yet none of us had seen her moving, though we were working at the goose-cache not more than twenty yards off.

Katrina's time was much divided between storing geese and attention to her baby. The baby, which very seldom cried, did so on this day at intervals. Very likely because, poor little wretch, the 'nyaninks' settled on its nose. Katrina danced it up and down in the cradle, consoling it with 'Pein-shaw, pein-shaw adski.' This word 'pein-shaw' they always used on these occasions. What it exactly means I do not know, but if you can give to it the same tone as an old nurse's 'There, there,' why then you have the thing exactly.

I came across a new flower in the bog to-day, a white ranunculus with a powerful scent, half-daphne and half-hawthorn. The idea of a *sweet* ranunculus pleased me so much that I kept stopping the sleighs and collecting it. I put some in my button-hole, where it smelt as strong almost as a violet. The Samoyeds were much amused at my delight over this flower. I had never even heard of a scented ranunculus, so no wonder I was pleased.

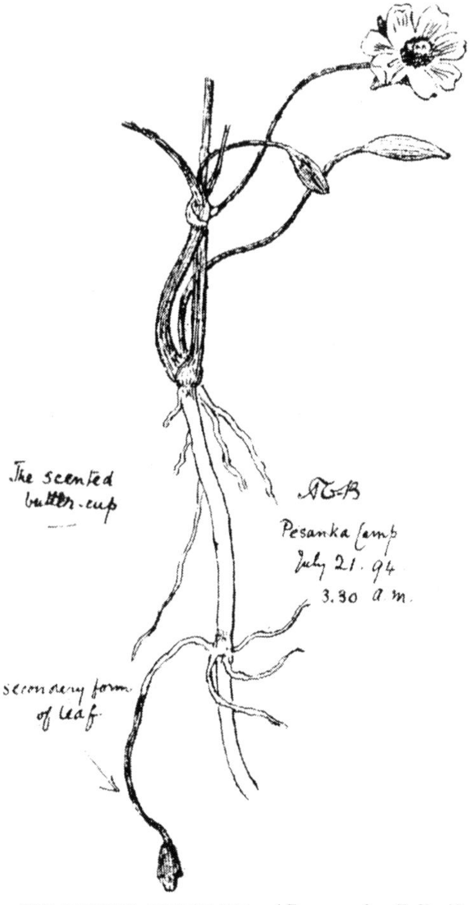

THE SCENTED BUTTERCUP. (*Ranunculus Pallasii*)

It was the only flower with a well-defined scent, except valerian, that I met with on the island.

About an hour before we were ready to leave old Marrk came up to me with five brent geese. I was sitting on a sleigh sketching, but out of the corner of my eye saw him coming up with a half-important, half-deprecatory air, followed by the eyes of all his choom. His lips were moving as he came, for he was rehearsing the terms of a little Russian speech, of which he at last delivered himself as follows: 'Yesterday no Marrk. Marrk bad yesterday. To-day Marrk good. Ahnglia five little geese. Little black goose very good. Yes, yes.'

His relief at getting this over was so great that before I could thank him fairly he had wheeled about and was on his sleigh. 'Now, all ready,' he shouted and started off. But the girls hallooed at him to 'stop.' It was anything but ready, for the geese were not covered up. So he came back looking very sheepish, among general laughter and jeers. I always fancied poor old Marrk was just a little bit touched in the head.

The Samoyeds distinguish the brent (*wurrah* in their tongue) as the 'little goose,' and the 'little black goose.'

July 20*th*.—After our return last night I improved my boat quarters very much by filling in the chinks with moss, so that neither the wind nor the prowling dogs could find an entry except by the front.

The heat to-day was really considerable, and the musquitoes and reindeer-fly very trying. I felt that I had regained the confidence I lost yesterday over the cap incident. I am sure I deserved it; for I worked like a slave at reindeer-catching and various forms of labour.

The reindeer when attacked by the fly behave very differently from cattle. Ordinarily, as I have said, they are scattered feeding all about, but now they draw up together to the accustomed place where the bucks are harnessed and picketed. The only explanation of this that I can see is that the uric acid which passes off the ground in such places is not agreeable to the fly. There they stand, heads down, feet stamping and hoofs clicking. Their heads are all one way, their tails to the wind. They leave an open avenue at intervals which runs the whole length of the herd. Into this avenue a beast will move when he can't restrain himself any longer, and then re-enter the herd at another point. I think this formation in columns is perhaps peculiar to the reindeer. Does the cariboo, I wonder, do the same? I wish some hunter would tell me.

I went nearly mad myself with the musquitoes while I was skinning my white brent.

One unpleasant result of the reindeer herd coming up so close was that the whole camp was smothered in hair which flies off the beasts with every twitch of their skins; and when the fly is about they are always twitching.

In the evening we moved off, going up with the idea of crossing the Pesanka and reaching a new goose-ground. Verrmyah had been off in the morning to visit his choom and get some necessaries.

And when we reached, after some hours, a hill some three miles from the river, we waited for him. Here I first gathered some plants of the pretty Arctic raspberry (*Rubus arcticus*). And here occurred one of those small simple scenes which well illustrate the happy, good-tempered character of these child-like people.

The girls were most anxious to get on and pitch the tents. 'It is no good,' said Uano, 'Pesanka deep. High tide.'

'It isn't,' says Ustynia. 'Is it, Yelisei?'

'No good; very deep; yes, yes,' says Yelisei.

ARCTIC RASPBERRY (*Rubus arcticus*)

Old Yelisei was so very old that in an ordinary way he was an ultimate appeal. But the girls were not to be done, for they had made up their minds, though all the men were against them with one voice.

'You men are cowards. We will go anyhow, and you will come, mother, won't you?'

'Yes, yes,' said Mrs. Uano, she would go.

'And you, Tinyah, you will come?'

The duchess didn't seem as if she much took to the idea of this deep water, but she couldn't stand out when all the women were going, so she said rather glumly, 'Yes, yes,' too.

'That's right,' laughs Ustynia, the ringleader. 'Now girls'; and in a second they had whipped up on to their sleighs and were off in a bunch, amid great laughter and shoutings from the men, and loud cries of 'Arkhoh, arkhoh'—'Deep, deep.'

This trivial incident is worth mentioning because it shows so well the absolute equality and freedom of the Samoyed women. Just fancy, if you can, an Indian squaw taking an independent line like this!

We waited perhaps an hour for Verrmyah, and then after a short journey suddenly there arose a roar of shouts and laughter from the men, for there were all the women on a little hill at whose foot the Pesanka ran. Of course they had not been able to cross. Only poor Ustynia, who had really tried the passage, was drying her things at the fire, for they were very wet.

Cub of blue fox — Nohò. Pesanka Goose-Camp.

Nothing could have been more good-humoured than the way the women took the joke which was now turned against themselves. I am not so sure about the duchess, who was fat, and I fancy was a bit disgusted at having been bustled for nothing.

I sketched a little 'psi-itz' or blue fox; and was much amused at the way it behaved to the dogs. They quite understood that they must not touch it. But old

Ni-arr-way would come up and stand proudly over it with his tail moving slowly, and very stiff, as big dogs are to a puppy. Then the little fox would lie over on its back and mew in a deprecatory way. Besides this it made two other sounds—a little snarling grunt for human beings, and the regular night-cry of the fox, not unlike the noise of a bantam who has just laid an egg.

July 21*st*.—I have not properly described the 'yier-serk.' It really consists of nothing better than some coloured cotton stuffs, obtained from the Russians in exchange for skins, and stretched over a frame of rough sticks. The wretched contrivance is far from waterproof. And as we had rain nearly all night, my poor friends were by the morning in a most deplorable plight. As for myself, I was comparatively comfortable under my boat, and during the night made more than one little journey round, listening to the cries of the foxes, and gathering mosses and plants. Finally, as the rain came on in torrents at about half-past eight, I turned in and went to sleep. About mid-day the rain lessened a bit, though the wind blew very strongly from the north-east. Then some of the men crept round a little, looking very miserable. It was useless, said Uano, to go to the goose-catching, for the boats could do no good with the wind from that quarter; and that there it might remain for several days. I knew from experience that this was true.

Now, though I was greatly tempted to remain and follow the goose-catching, two considerations influenced me in the contrary direction. In the first place I was not easy about Hyland; I was afraid he might be having one of his attacks, and I knew too that he would be wanting bread. Nor again could I feel that I had a right to amuse myself until I had done all that lay in my power to help the *Saxon* should she come. So, very reluctantly I determined to return to-day.

Before I left, Mrs. Uano gave me three small loaves of bread, and a little flour with which to thicken our soup.

I left at six in the evening, and Uano came with me on a second sleigh.

He surprised me very much by the way, for he asked me to look through my glass at a white object in the distance and tell him whether it was not a reindeer. I knew perfectly well that it was a snowy owl, for I had been watching it for some little time. And a snowy owl it was. But it seemed odd that a pair of native eyes should have been so deceived.

After crossing the Baroshika, which was again very deep, we stopped to bait the deer. It took them exactly ten minutes, I found, to clear the lichen which lay around them without moving from where they stood.

When you want the deer to eat their fill, you lay the 'toorr' down in front, as I have before described, and then, when they have fed up to it, move it farther away, and so keep moving it each time as they draw up.

The amount reindeer drink at this time of year, while they are growing new horns and shedding their coats, is perfectly astonishing. The whole way along one or the other is either trying to drink or else is snatching mouthfuls of snow.

About midnight we reached Scharok. Hyland had just finished his supper. He was very well.

'Well, sir,' says Hyland, 'I *am* glad to see you back. It's pretty lonely down here by yourself, especially with these divers always screaming.'

'They certainly are the most melancholy birds,' I replied. The red-throated diver, you must understand, has a cry much like a woman's wailing.

In addition to a female Siberian herring-gull, a pair of glaucous gulls, and a female snowy owl, Hyland had obtained, while I was away, a sanderling in summer plumage.

I had brought with me four geese, which we boiled, and Uano and I had supper together. Then at 2 A.M. he jumped on his sleigh and galloped off across the creeks.

Hyland had one bad little bit of news for me, for the turnstone had escaped and the waiter had died.

There had come an evening when the fog fell heavily and threatened cold. Yet the waiter was making a good supper off his store of equisetum when last seen by Hyland. But the fog changed to frost; there was no warm wing of a mother to run under, and I suppose that did it—the cold.

For the morning was white with frost of an hour when the sun looked into the kraal of moss. And there on his back lay the poor little waiter dead—there was no doubt of it—dead.

CHAPTER III

SCHAROK AGAIN

Sunday, July 22nd.—Soon after Uano had left us last night, or rather this morning, the wind got up very wildly from the north-east, and was accompanied by a heavy drizzle of fog. As I did not feel sleepy I sent Hyland to bed, and then went for a two hours' stroll down the coast. The ice looked more promising, for it had moved out to sea, and between it and the sand-banks was quite a wide bit of open water.

It was not a cheerful walk, and at five o'clock I was back and ready to sleep.

I was up again at six, and after an eight o'clock breakfast went to sleep again till eleven, for it was such a wretched day that this seemed the best thing to be done. There was a good deal of skinning and writing to be finished, so we did not have luncheon till 6 P.M., but then, as it was Sunday, we finished the last of our little tin of dried vegetables, which we had been eating once a week.

Hyland told me the following interesting story:—

'The day before yesterday I was up by the lake when a cock snowy owl came by and began cutting down at the young of a pair of white-fronted geese. Both the old birds protected their young. The gander flew right at the snowy.'

'How do you mean *at* the snowy?' I asked, anxious to get the facts correctly.

'Well, sir,' says Hyland, 'he rose clean off the water and struck at the owl with his wings. After this the owl flew off.'

Although it seems that the snowy does not nest on our island, they come there much on migration; and a day scarcely passed now without our seeing one or more of these fine birds sitting on a hummock of the peat. As there are no lemmings or other small mammals here, it follows that the bird is obliged to make its dinner of fish and birds.

In the evening I went for a long walk round by the lakes, and returned in a dense fog.

We found our first young Lapland bunting to-day. I lay up for a long time by the big lake looking at the red-throated divers. There were seven of these birds on the lake, and they were all at play.

We were now growing rather anxious about our water-supply. For the snow from which we got it was rapidly going.

July 23rd.—Another horrid day. It was not in reality so very cold, for the thermometer rose at noon to 44° in the sun. But a gale from the north-east and a soaking rain made it most unpleasant. Nothing could be gained by going out, for you could only struggle miserably along half-blinded. Hyland complained of a bad pain in his back. 'I'm all right when I lie still,' he said, 'but whenever I move it shoots all up my back.'

It did not seem to be rheumatism exactly, but still the tent was soaking wet and everything inside it. So I made him move up into Uano's hut, which was, at all events, more or less rain-proof, while I did the cooking for the day. The white-fronted goose came as a pleasant change from long-tailed duck, on which we had chiefly lived.

For the first time I noticed at midnight the sun's lower limb below the horizon.

'The ice is all back again, packed right against the land.'

July 24th.—With a change of wind to-day to the south-west and then to north-west, the ice began cannonading as usual.

Found all the biscuits wet, so turned them out to dry. 'We have now exactly forty-five of them left, which puts us on an allowance of three a day for a week, by the end of which time the *Saxon* or else the Russians may have come.'

The snow-water had failed us. But we found a hole filled by yellow water from the bogs, and down to this we moved the tent.

The male snow-buntings were now changing to autumn plumage.

July 25th.—I had become tired of this inaction. The time was slipping on, and as yet we had done so little. So after breakfast (fried brent, salt pork—the remains of that we gave the Samoyeds—soup of grey plover, long-tailed duck and sanderling) I made up my mind we must patch up the old boat and go out that we might sound the channel and make ready for the *Saxon.*

The wretched old craft leaked so badly that it took a very long time to plug and patch her up. And even when that was done we were in much the same fix as Robinson Crusoe with his fine new boat. For she was lying high and dry among the grasses. However, we stuck to it. Our great want was rollers. There was nothing for it but to use the wood of some of the fallen crosses—we could not help ourselves. With these and an old oar or two, gradually and with infinite exertions we worked the boat inch by inch over the snow and mud (into which we sank deep at every step, and in which we had to probe for our rollers) till we had it at last by the water's edge.

We put our guns, some cord, a broken kedge, and old Sailor into the boat, and finally got under weigh at half-past ten at night. The tide was running strongly out—a three-quarter ebb—and the moon was in her last quarter.

The tide was running out so fast that Hyland's chief

concern was to keep the boat's head straight, I steering and sounding along with a hammer tied to a cord.

The depth, on the whole, varied from six to ten feet, but even at that was crossed by a shoal in two places; and here, before we were aware, we ran aground. I got out, hauled the boat off, and then waded on for some distance towing the boat and sounding with a pole.

The main channel after two sudden turns opens on the entry by a deep reach. On the southern side of this entry we landed on the outer sand-banks.

Here we succeeded in hauling the boat up to the very top of a high sand-bank, where it would lie quite safe in any event but a north-east gale.

This is that outer ridge of sand which separates the mud-flats from the sea.

We found the ice grounded some half a mile to seaward, and thence stretching away as far as the eye could see.

I walked off to inspect a big lump which stood up from the sand about two and a half feet. I found that it was deliberately built up of weeds[1] and hydrozoans, round which sand had collected, and in the hollowed top was a single egg of the glaucous gull. The pair of these birds were flying over my head.

We walked along the ridge. We came upon a whole row of glaucous gulls' nests, but none of them held any

[1] They included the following species:—
Fucus vesiculosus, *Fucus serratus*, *Ceramium rubrum*, *Corallina officinalis*.

eggs, though all had been newly made up. There were many old sleigh tracks about the sand to show the reason why young gulls of this species were so rare at Scharok.

Three handsome goosanders were fishing off the bank.

July 26th.—We got back to the boat, which we found well out of reach of the tide, at 3.30 A.M., and collecting a few little sticks and bits of dry seaweed, managed to make a tiny fire with which we warmed some Liebig.

Then with sunken logs we improvised some extra anchors for the boat, and with food, guns, and old Sailor, travelled south once more for further explorations.

The sun was brilliant, and the glaucous gulls looked lovely as they sat about the nests. One Arctic morning such as this repays you well for the gloom and unkindness of many days.

Two miles along the bank we found a large pine-tree drifted. Here we made a fine fire and went to sleep. Soon after 10 A.M. I woke to find a chill fog and the sun fighting it. I roused Hyland, and while he made up the fire and warmed some brent goose, I walked back again to see to the boat; for I had been fearing she might have been lifted by the tide. But she was safe.

A large flock of common scoters were diving in the tideway. They preserved the most perfect order. First the leading bird dived, then the next, and so on till all were under water. Then rising they would all crowd up together, and again string out and dive as before.

Many medusæ had been left by the tide, and at the head of a small inlet some mollusca[1] were very abundant.

The boat I found quite safe, although the tide had been unusually high. Now, I thought, we can walk down until we reach the Pugrinoy on the gulf. Not so; after another mile or so of the sand-banks we were suddenly brought up by an absolutely impassable river estuary.

But we followed up the bank by its various windings, until at last, in spite of very deep mud, we were again on the mainland.

Here I found a reeve with two little downy young ones.

So I walked on to the south-west until all of a sudden we spied three sleighs coming over the tundra from the north. We sat on a hummock and waited.

The sleighs hesitated and stopped short. After a pause one of them came on alone. It arrived—our dirty little Shabla driving four reindeer cows.

He was far more pleased to see us than I to see him, for I had never cared much for this tiresome, sly little

[1] These included the following widely distributed Northern forms, viz.:—

GASTROPODA—*Chrysodomus despectus* (Linn.).
Buccinum scalariforme Beck.
Natica grœnlandica Beck.
Margarita undulata Sowerby.
PELECYPODA—*Cardium grœnlandicum* Chemn.
Mya truncata Linn.
Macoma balthica (Linn.).
Mytilus edulis Linn.

man. Seeing him in conversation with us, and taking no harm, his companions started also to come up.

It appeared that Shabla had paid us a visit at Scharok, had found the camp deserted and the boat gone. This was the way in which he told us about it :—

'Shabla to-day Scharok. Sun so,' pointing to the sky to show the sun's position at that time. This was the regular way of the Samoyed when trying to explain the hour. 'Choom there shut. Ahnglia? No. Hylum? No—Where? Then see boat—arnoh—not there. No, no. Boat where? Not good. Much ice. Go see. Yes, yes.'

The two new men were the brothers Bulchikoff. Dirty, good-looking, silent men; very proud to be sketched, and posing themselves in an amusing way on hummocks for that purpose. They said their camp was on the lower Gobista. Old Marrk, they told me, had the upper waters, and I found out later this was so.

So I told Shabla I was walking to the Pugrinoy. 'Pugrinoy,' says Shabla; 'no, no. Bad, bad. Much river. Much marshes. No walk; soon go boat; no ice.' He was dead set against us trying it. And indeed there seemed much force in his remarks.

The dirty brothers Bulchikoff gave us a bean goose, of which they had five on the sleigh. Then they rounded about and were off at a gallop, Shabla's little cows[1] getting over the ground at a capital rate.

[1] See note on Mammalia.

We were back at our fire on the sands by 11 P.M., and after eating the bean goose lay down to rest.

July 27th.—I woke at 3.30 A.M. to find a north-east wind and freezing fog. After a good long walk along the sands, during which I saw many brent and many sanderlings, I returned at seven and woke Hyland, and back we walked to the boat.

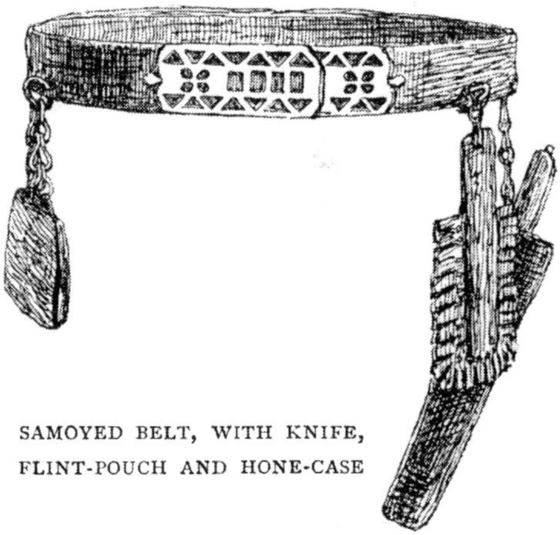

SAMOYED BELT, WITH KNIFE, FLINT-POUCH AND HONE-CASE

We had anything but an easy time of it getting home; and had I not recorded the bearings very carefully when we came out we should have got into a mess. Any one who has tried to navigate the creeks in an unknown estuary with tide at ebb knows what that means.

On our return I reproached myself much for having taken Hyland out when I heard him say, 'Well, sir, I do feel bad. My back is all aching, and the pain shoots all about.' I imagined it was liver, so sent him to bed with physic. At 4 P.M. I went up to Hyland's hut. He was fast asleep. When I visited him again at 6 P.M. he woke up at my entering, and said he was worse.

However, I made him eat when I had cooked a brent

goose and obtained some first-rate soup. At 10.30 P.M. I left him in his hut fast asleep.

That night I had to write my diary by the light of my little pocket lantern.

It is easy to see from the observations we had made that the *Saxon,* drawing her twelve feet of water, could not hope to enter Scharok harbour. But at any rate her boats could come in well enough if the ice allowed, and the next step was to mark out the channels for her in some way or other.

July 28th.—So at 1.30 this morning, just as I had finished my writing, instead of turning in to sleep as a wise man would have done, I must needs go walking out to consider this question.

The tide had all but run out, the little creeks were sinking out of sight one by one, and soon there would be nothing south of our channel between the mainland and the outer sand-banks but two miles or so of black ooze. It seemed to be my chance.

I set to work. I carried some beams down from the huts, and falling to with a will, chipped away with the axe till I had reduced them sufficiently for my purpose.

First I drove one into the ground opposite that point where the main channel, after coming straight in from the flats, turns suddenly north to run under our camp. Then I set out bearing the axe and a couple of long posts across the mud.

A little rhyme kept running in my head. 'De blind hoss stickin' in a big mud-hole,' it began.

I admit it was not a very wise undertaking without mud-pattens, and more than once I nearly paid for my rashness. However, I went staggering along with my load, drawing my feet out of the treacherous mud as best I could, and counting every pace. I knew I had time enough for my job before the tide turned if only I could keep myself from sinking. A heavy man would not have stood a chance, and as it was, the load I was carrying handicapped me much. Once I was all but checkmated. For I came suddenly to a creek into the edge of which I sank above the knees, and seemed to be going altogether. But pitching my posts down on the firmer mud, I managed to scramble out on them, but what with the scrambling, and what with the fishing for the posts again in the black, smelling mud, and the loading them on my shoulders, I was a most unpleasant body. But I stuck to it, crossing the creek a long way up, and coming down its further bank to the same spot to take up the record of my paces.

At last I reached the first point where the creek turns to the south. Here I drove in one of my tall posts, which I judged would show some three feet for guidance at high water. Of course I couldn't hammer it in straight, but I got over this difficulty by driving it in slantingly, and then shoving it up till it stood straight and fairly firm, the foot well blocked about with mud. It would stand, I judged, in spite of the tide, long enough for my purpose. This was just five hundred paces from the start.

Then I pegged away to another point eleven hundred paces off and did the same thing.

The whole distance was not very great, but when I turned to go back the tent looked a great way off, in the knowledge of the mud I had to re-cross.

Relieved of my load, I went more quickly, but also with less care. And twice I was caught in a sort of mud quicksand, and got out by a narrow squeak, my feet being sucked back with each attempt to extricate them. Then the rhyme ran quicker than ever, 'De blind hoss stickin' in a big (suck) mud (plunge) hole—(suck again, and at last) Doodah, doodah,—dey!' I was out. I laughed as I thought what a very blind hoss, not to say an idiot, any one would think me who could have seen me then. It was a poor business, but I got back all right, and after proudly surveying my handiwork went to bed.

I was up again at five o'clock this morning, and went to the hut to see how Hyland fared. He was sleeping heavily, and I left him so. The day did not look very inviting, with a thick white fog and north-east wind; so I went to sleep again.

I paid Hyland several visits during the morning, but always he was the same, scarcely able to move, and complaining bitterly of acute pains across the loins.

This was the day on which the *Saxon* was due, but obviously she could not have got in. I kept an eye to seaward all the same, for how wide our ice-belt actually was we did not know.

But there was never a sign of a sail nor a steamer's smoke on the horizon.

We were now fairly settled down to the full idea of camp life, and although I never ceased to regret the want of many scientific necessaries which had gone away with the yacht, still as regards the means of existence we were fairly well supplied. It is true that the food we had brought with us was all but exhausted, and though we managed to make a few things—our ham, for instance—struggle on for a bit longer, this was only by not eating them, or by taking but an odd mouthful in the day. But we could always shoot a long-tailed duck, and though we didn't like it, it proved sufficient. We had wood, and though it was not easier but harder to cut up than before, for we had long since burnt all the easy pieces, we were better hands now with the axe. We had water, though we had to boil every drop before drinking. Even dear old Sailor was comfortably housed; he now lived in the barrel which once had held our food, rat-traps, and other appliances. My capital rat- and mouse-traps were useless—there was nothing to catch.

By this time we had proved to demonstration the absence of all native terrestrial mammals except the red and the blue fox. Not a hare, not a lemming lived upon the island.

But we badly wanted an oven; something that would economise our fuel and keep the pot from the cold, which made the meat take far too long in the cooking. I had quite determined that an oven we must have.

Now there were lying out on the mud some big slabs of sandstone and conglomerate. I put on a pair of india-rubber long-boots and went for these. Those I wanted most I could not lift, but the next in size I tackled, and after incredible back-breaking exertions, managed to get them out on to the bank. Once the mud pulled off my right boot, and while I was hopping round on the left leg trying to keep the right foot dry, I lost my balance and—squish squash—in went the right foot and down, down, sending up a squirt of liquid mud and smelling bubbles. You would have laughed if you had seen it. I didn't.

Inch by inch I got my big stones up the bank, and built in a hollow quite a first-rate oven. At any rate, it would take our only pot and the little kettle.

I cleaned and cut up some duck, stuffed the pot full, lit the fire, and was delighted to find it roar. I was squatting by it rejoicing in the white heat that showed inside, when it suddenly went off like a cannon, blowing the pot out to the entrance and smothering me with bits of stone and ash. A big stone had flown to bits. However, I got another, and by heating it more gradually all went well, and we had no further blastings.

I made a little geological excursion after that, hunting up a river for sections and fossils, but that was all I could do. The wind had gone to the east; it was drizzling and blowing half a gale, and at high tide the ice for the first time came right in at the harbour mouth till it grounded all about on the mud, the biggest bits floating up the channel under the tent.

Sunday, July 29th.—Any one would have thought that we might have rested content with having once taken out a half-rotten boat and come back in safety. But we were hungry men. There was only one wretched long-tailed duck in the larder; we wanted geese, and geese we saw on the sand-banks away out on the ice.

'What a pity we can't get one of those old geese!' I said to Hyland. Esculapius could not have done it more quickly! He had been ill and unable to move beyond a crawl till then.

'We can get one, can't we, sir? I can pull the boat out.'

I knew very well he couldn't, with his bad back, but the tide was running out, so it was only a question of keeping the boat straight. I thought I could steer her through the ice.

Well, we started. The ice did not look very pleasant at times; it could easily have knocked a hole in our ancient craft. Some of it was in very pretty, light-topped pieces ten feet or so above the water, with hollows and green caves in them, and gulls sitting on the spires.

We managed to dodge it, and landed near the harbour entry, but at such a steep spot that we couldn't get the boat up. So I left Hyland there to fend off the ice while I took a walk.

I could not come at any geese, and after a long walk returned to find the tide just making. Such a body of ice was coming up that I really felt it was not wise to risk it. The boat would move in a bit out of the ice's reach as the water lifted; so I concluded that I had better run home and bring some food.

I set off carrying gun, axe, and camera, tried to cross a creek three times and failed, ran and splashed and tumbled through the mud, perspiring for the first time ('I have it in my notes') since we landed on Kolguev. Finally I waded across, and walking three miles up the coast reached camp, cooked long-tailed duck, and left again at midnight with my companion's food.

But I had not gone two miles before I met him coming wearily along to meet me. The tide had lifted the boat, and he had been able to haul it up out of harm's way. Then he had walked south along the sand-bank and had crossed at the head of the creek I had waded.

The glaucous gulls, he said, had mobbed him most persistently, actually touching his cap. He was afraid they would strike and hurt him, so he had to shoot two in self-defence.

And so ended our second boat adventure. But the boat was now far away on the sand-banks.

July 30th.—No sooner were we back in camp this morning early than the rain came on in torrents, and a gale sprang up from the north-east. The rain held on till three in the afternoon. Then we went out to the lake, and shot an eider duck that was there with her little young ones. We also picked up a well-fledged little stint.

On the way home we shot two willow-grouse out of a pack of eight cocks. The first pack we have seen.

The rain came on again in the evening, and my tent was flooded as usual.

We were just cooking supper at 10.30 when up came Uano and young Yelisei, bringing me my geese; forty-six brent and three bean.

Then they came and waited by the tent, in the silent way that many natives have, waiting for us to open a conversation.

I pointed out to him the boat away on the bank beyond the creek. Such a tempest was raging, and the water broke over the ice so furiously that at that distance it looked as though the boat must be lost.

'Propalo' (spoilt or lost), says Uano.

This Russian word did duty with the Samoyeds for many ideas. A dead dog or deer was 'propalo,' a fly-blown goose was 'propalo'; a mislaid axe, a rotten cord, a worn-out coat,—these things were 'propalo.' Only a dead man was not 'propalo'; for him they had another word, a word of their own, which fact perhaps pointed to a belief that he was still going on somewhere.

'No, Uano, not propalo. Very well,' say I. For I had seen with the glass that the boat was well protected by a barrier of grounded ice.

So then we cooked a goose. Our friends were amazed at and enraptured with our oven.

'Ah,' they said, 'you all stone. Yes, yes. House stone, trail stone, fire stone. You very good. Yes, yes.' For we had astonished them much by describing our English stone houses and roads.

The Samoyed word for 'yes, yes' seemed always to me the best affirmative I knew. The Russian 'da, da,' or 'yah,' or 'oui,' or our own 'yes,' none of them seemed so clinching and so final as the Samoyed 'drem, drem,' said as they could say it.

So they ate the goose: and when that was finished I made them as dry a seat as the wet tent permitted, and gave the old man a pipe, which was always his great joy.

And then, 'Uano,' said I, 'what news?' For I knew perfectly well from his face and manner that he had not come only to bring the geese.

'Ahnglia, Hylum, stay with Uano all winter. Yes, yes—choom,' said he.

'No, no, Uano, steamboat come, or Russians come.'

'Steamboat no, no, steamboat gone away,' he said. 'Yes, yes, far away, all ice. Yes, yes'; and he waved his hand across the sea.

There was something more in this, and after much patience I wormed it out of him. It appeared that five days ago—that would be Saturday, the 26th instant—he had seen a steamer off the northern end of the upper sandbanks. She was there, he said, all the morning, but in the afternoon was gone. Of course he had fully made up his mind that she was our boat, and as he always hoped we should be with him all the winter, so now he felt that sure.

But I was of a different mind. The date was too early, and I knew quite well that if the *Saxon* had succeeded in coming so close as that, she would not have gone off so quickly.

Now came the question, what were we to do? We had no vegetables, no bread, and only a few biscuits in the bag. I attributed much of Hyland's aches and pains to want of vegetable food. Round our present camp all was bog or peat; but up with Uano there were grassy banks where sorrel grew. I knew also that from Pesanka camp we could keep an eye on the sea. So I told the old man that he must come down to-morrow and bring us sleighs so that we could return with him. Of course he was delighted.

July 31st.—So we sat there talking and talking, in the hope that the wind would fall or the rain stop.

Is it in Dean Ramsay, or where? I forget. But the rector says to the old clerk—

'This is getting very serious, John. We must have on Sunday the prayers appointed for rain.'

'Yes, sir, very good,' says old John; 'prayers 'pinted.

But, bless you, sir, 'taint no manner of use so long as this east wind lasts.'

And our rain wouldn't stop. So at 2.30 A.M. the Samoyeds went off in a real pelter. We had never seen the deer travel so fast before. For now they were clean moulted. I must confess I envied Hyland his hut. It is true that the wet would trickle down the walls in places, but I was flooded. There was just one dry corner in the tent, and here I curled up. But sleep was out of the question. Every moment I thought the tent must go. But it stood wonderfully. That is the best point about these 'Whymper' tents, and a very strong point it is. But to pretend that they are waterproof is not even reasonable humour.

The rain did stop at five o'clock. I contrived a gutter to carry off some of the water. Immense flocks of gulls had come in and were sitting out on the bog at the back.

I wrote a note explaining to Powys our movements. This I put in the chapel in a most conspicuous place.

'We are moving up about twenty-eight versts,' I said. 'We shall be in full view of the sea. I shall send a team down here every other day. The *Saxon* cannot get in, but either boat can enter well. Sound carefully. Keep the two posts you will see on the land in line for three hundred yards. Then five hundred yards N.NW. by compass (be careful here of a shoal and a sunken headland). Then, bearing due west, you can reach right in to the post one hundred yards below the southernmost hut. Be careful to avoid the first creek on the north as you enter. It is a *cul de sac*. You will do it best at three-quarter flow. If you see nothing of us, keep the steam-whistle going, and make plenty of smoke.'

The ice had swept away the posts I had been at so much pains to fix up. This was the best I could do; and it seemed not impossible that the boat might land higher up, and Powys walk down to the huts, where he would find the note.

The Samoyeds, Uano and Onaska the Prophet, turned up with added sleighs at six in the evening. We sent them down to bring back the old boat while we packed up. The geese we cached in the Samoyed manner, and all was ready.

Everything we had was soaking wet, and the wind and rain could only fitly be described as raging.

We reached the Pesanka camp at midnight. And so for the present we had done with Scharok.

HEADS OF SAMOYED WOMEN
Wrapped round with two of my 'present handkerchiefs

PART V

Peaceably and with little event our life with the natives goes on—Reindeer are killed and wild swans are shot—We have a visitor from the mainland of Arctic Russia

CHAPTER I

DAYS IN CAMP

August 1st.—The month opened miserably, for a perfect hurricane raged, accompanied by torrents of rain. We indeed managed to get the tent up with all hands at the work, and extra guy-ropes to keep it standing. Even so I had to prop it up from within, the wind pressure was so tremendous.

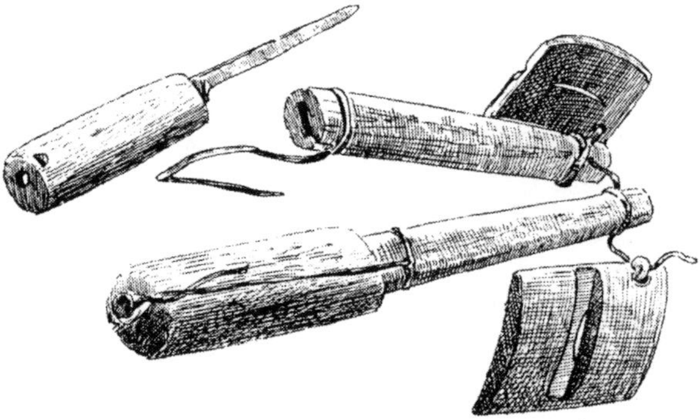

No one ventured out; our whole time was taken up in fighting the wind and water.

This went on till five in the evening, when in a lull of the storm out we crept like half-drowned rats, to shoot with bows and arrows, and throw the di-zha at a mark.

Samoyed bows are very interesting. Like every other

wooden thing these people have, they are made of drift-wood. A special plane is used for that purpose, which the figure will help you to understand. The wooden gauge has a slit into which the triangular knife is fitted. This bow belongs to those known as reversible bows, or bows which turn inside out. But it seems to be peculiar in this, namely, that it is a simple and not a compound bow. That is to say, it is neither spliced, nor whipped round with string or skin. None the less, it retains its elasticity in a remarkable way. But always before it is used it is dipped in water, and then rubbed with the hands, which gives it suppleness and spring.

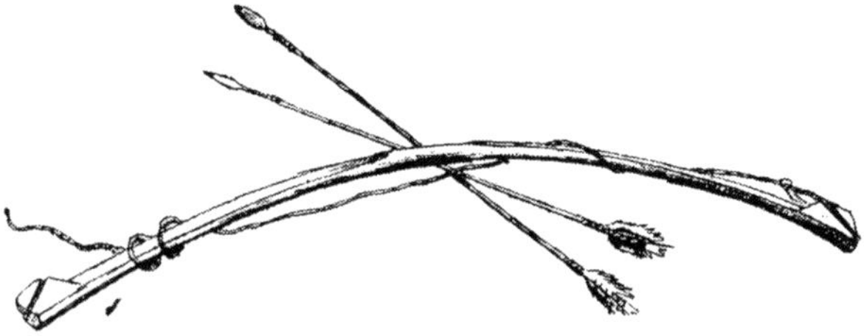

We fixed up a bit of stick for a mark, and began to shoot at fifteen paces—Kallina, Marrk, Mekolka, Hyland, and I. It was a competition.

The men threw themselves into a curious attitude when shooting. They used the bow horizontally, not perpendicularly as we at home. As they pulled they rose on tiptoe and then almost crouched till the bow was on a level with the eye. We held our own pretty well at this game, but Mekolka was champion; he was always best at everything. The peg was removed stage

A REINDEER HELD WITH THE DI-ZHA

Face page 273

by stage, until at last it was about forty paces away. At this distance Mekolka made very fair practice. Whether the peg was hit or no, his arrows were nearly always sufficiently near to have transfixed a goose had one been there. Of course it was absolutely nothing in the way of archery; but one could quite understand how a prowling Samoyed could creep upon his quarry, and secure it by these simple means.

Then we threw the di-zha. They were far and away better than we were at this; and shouts of laughter greeted my feeble efforts to entangle the stump. Nothing could exceed the patience with which they tried to teach the proper method; the gathering it up in the hand; the swing, the throw, the drawing tight. Just as a cowboy at the moment of noosing a bullock turns his broncho to resist the strain, so in many cases the Samoyed turns his back to the struggling deer, bringing the di-zha across his thigh. But not when he has a very wild or strong beast to tackle. In this case he yields very gradually, contesting every inch and letting out the six or seven yards of the spare end of the di-zha, which always trails on the ground, until one of his fellows comes to help.

At last I caught the stump.

'Good, very good, yes, yes,' cried Mekolka; 'Hylum, Ahnglia soon very good. All winter in choom on Kolguev. Hylum, Ahnglia soon Samoyed. Yes, yes.'

'No, Mekolka,' I laughed; 'soon leave Kolguev. Go with Russians; England very soon now.' At which

Mekolka looked quite sad. For they are a simple folk, and make no pretence of concealing their emotions; and Mekolka had quite made up his mind that they would have us with them all the winter. But now he returned to the attack.

'Kolguev winter very good. Yes, yes very, very good. Walrus—bear. Samoyed shoot bear. Pesanka mountains. Mekolka go his sleigh quick, shoot many bullets. Bear dead quick. Yes, yes. Big moon.'

The Samoyeds did their best to prevent their wretched tents from blowing away by piling boats and sleighs round them. Also we had a boat behind our tent, which broke the wind a little, or I think it must have gone.

Onaska went down to the creek, and brought away two sleigh-loads of brent from the cache, which our friends set to work to pluck.

SAMOYED WOMAN'S WORK-BAG

August 2nd.—The gale, which raged all night, fell off a bit during this day, and Verrmyah drove away to get his choom. With this he returned in the evening, and the whole Uano family moved into it.

Kallina came and sat in my tent door to make his arrows. The shaft is of drift-wood winged with grey goose feathers, split in half, but not trimmed. The heads he worked out of old knife-blades, old files, and bone. These points are not barbed. The whipping of the feathers and heads was partly of sinew-thread, partly of split quill. Kallina took a great fancy to my butcher's knife.

'Very good knife, yes, yes. My knife now; Kallina knife,' he said. These people have a sudden light-hearted final way of appropriating your property just as children have.

'No, Kallina, my knife.'

'Oh!' said Kallina so sorrowfully, and with such a world of disappointment in his voice, that I could not choose but say, 'Yes, yes; when my ship comes, then Kallina knife.'

You should just have seen his delight. He stroked the knife, and ran his fingers deftly along the edge; he came and shook me by the hand to seal the compact; the knife was already his, and he was skinning his seals by anticipation.

When I went for a long walk in the evening with old Sailor, I found the yellow Tromsö viola out (*V. biflora*), and that geum or water-avens (*G. rivale*) which grows on our Hampshire water meadows, and is so common on the Solovetskii Islands. But the common *Artemisia*, which was in bud a fortnight and more ago, was not yet out in flower.

All the Yelisei family were hard at work preparing geese for the barrels. It was a fair sub-division of labour. The old man led off by splitting them with his axe, the women and elder girls gutted them; then they threw them on to the children, who took out the gizzards and chopped off the beaks. These were carefully preserved threaded on strings. Some were then used as necklaces and ornaments for the hair; some as beads for the little dolls, or, as they call them, 'ooquoh.' These are made to imitate fathers and mothers of a family, sons, babies, and so on.

Why does Nordenskiöld put these Samoyeds at the *bottom* of the Arctic Mongol group? He makes them the lowest. And why does Carlyle, casting about for an instance of hopeless barbarity, pitch on the poor Samoyed? This book, if it shows anything, will surely reveal the Samoyed as an extremely intelligent man, far and away more so than the Red Indian.

Take Mekolka. Mekolka had been born and bred on Kolguev. He was now about seventeen, and until last year had not known how to read or write. Then with the Russians in the summer crossed the Prophet, Onaska, who had learnt in the Archangel school. Through the winter he taught the boy Mekolka, and now this lad could not only spell out Russian slowly, but write quite fairly well. His father, who had always used the notched stick for his calculations, now trusted much to Mekolka's powers.

I used sometimes to have a sense (which I never had with the Indians) that these were, so to say, our own natives, just English yeomen families, or so. The very laughter of the children seemed to have an English ring, and the tone of their voices was not very different as it came down to us from their choom. And often as I crept round the camp of an evening and lifted the flap of their choom, I used to ask myself:—'Now, how are these people in any way essentially different from ourselves?'

Softly came the voice of Katrina crooning to her baby. She had a little song for it, just as any mother here will have. 'Adski, adski, pi-sing-a.[1] Adski, arki hu-wi-ni-ti-ân,'[2] this is what she sang. I don't think it sounds very pretty to you, but it means this: 'Laugh, baby, laugh. Baby will soon be a big man.' And then she would play hide-and-seek—'hunyan hordah' she called it, covering its face with a bit of soft skin. 'Hunyan,' where is it? and 'hordah,' found.

And the children in the corner would be having a tea-party for the dolls. Nyabkutni, the curly-headed, would work the hosts, the father, mother, and their family, and the other girls would introduce the guests. The choom they represented by three sticks. They used to go through most elaborate performances, and shout with laughter, but I could never gather a single word from all their chattering, and whenever I asked them, they only laughed the more.

Hyland was very wretched yesterday, shivering and ague-like, and did not move all day, so this morning when I went out about five o'clock I did not disturb him.

[1] Laughter. [2] The day after to-morrow.

August 3rd.—When I came back at 8.30 he was still sleeping heavily; he had been to sleep so many hours that I thought I ought to wake him up; but as he still complained of feeling very unwell, and was asleep again in a moment, I let him lie. Later on in the day I

THE HEADS OF THESE DOLLS ARE MADE OF THE BEAKS OF BRENT AND WHITE-FRONTED GEESE

persuaded him to take a walk, and after an hour's stroll he came back feeling better.

To-day we caught and killed a young reindeer. Mekolka went out 'adliurgo' with the dogs, and brought the herd up. The herd was then gradually divided, so that between the two halves stretched the width of the sandy platform on which we were camped. Then, while the mother and calves galloped across from one side to the other, Uano pronounced which calf he would have caught. Immediately all eyes were fixed on the victim. It was a fine grey-coloured calf, and ran nobly with his head well up.

Then every man gathered up his di-zha and crept away, half-crouching, like a stage conspirator. It was very difficult to take this calf, because he kept in the thick of the crowd, and more than once the wrong animal was caught and then set free.

At last Mekolka, creeping behind a group of bucks, with a sudden rush and a toss of the di-zha had the calf by the horns, for he had velvety horns some foot in length. Even then he took a lot of holding, but a second di-zha was thrown round him, and the men advanced on him, hand over hand. At once he was hit between the eyes with an axe-head and stunned. Then, with a rapid movement of the knife, the axis and atlas vertebræ were separated, and the poor creature's heart stuck under the lifted leg, so that no blood escaped.

After the women had flayed him, which was the work of a few minutes only, the ribs were cut away, and then every person in the camp came round to a dinner party of raw reindeer.

I can't say I admired the performance, though it would make a very striking picture done in colour. You can fill it in for yourselves. For scene you have a raised plateau of the tundra, a line of red-and-blue cotton gipsy tents, a birch-bark covered choom, a troop of quarrelling dogs, a confusion of boats, sleighs, tubs, and nameless paraphernalia, a herd of deer scattered round over the waste, and a team or two of five fastened up. Put in the middle a circle of men, women, girls and boys, standing or squatting round the dead body of a fresh-killed smoking deer, cutting, tearing and feeding, and then you have all.

They were astonished beyond measure that I did not join them. Hyland ate some and pronounced it excellent, and I really don't know why it should not be so; but I prefer my food cooked. The Samoyeds were especially fond of the windpipe, the sinews, the lungs, the outside of the bones, and the velvety vascular part of the horns. All these they dipped in the blood before eating.

'Good, very good,' says old Ustynia, offering me a most tempting bit of lights, fresh from the blood. I made a face which expressed disgust, but they only thought it meant that I wasn't hungry. 'Oh, not hungry,'[1] they said; and 'Oh, not hungry,' went the round of the ring, in a series of intermittent ejaculations between the mouthfuls. They were hungry, it seemed, for they gorged away. If any one in this country could have eaten so much at a go he would have fallen asleep. But

[1] Yangho ormüngawum.

they danced—a silly dance, with a song to it, 'like a broken-backed Gregorian,' as I said in my notes at the time.

Then we played 'meitz hansànquoh' (*meitz*, a ball), which is rounders, or baseball, with slight differences. We saw no other outdoor game. Pretty nimble they were on their little bowed legs, though of course we could run away from, and all around them. Hyland recovered in time for the game, and played very well indeed, wearing out his sealskin 'pimüh' in his efforts. He was a much better shot with the ball than they, and though he was so big they very seldom hit him.

In the evening the wind dropped, after eighty-four hours of gale.

August 4th.—A warm day; thermometer up to 61° in the sun; very calm, very pleasant.

The Yelisei family packed up to leave, the eldest Yelisei girl, who was nice-looking, but very silent, excepting in her circle, had such a fine head-dress, with a beautiful native-made star on the forehead, and many rows of beads in the hair, that I wanted to buy it. I hadn't the courage to ask herself, but tried to explain to her brother. This was much worse, for he immediately concluded I wanted to marry her, and in a moment I was the centre of a delighted group. They all chattered at once, and all pointed from me to her, and from her to me. The poor girl—I cannot remember her name, and don't think I heard it—blushed, and looked so confused that I was quite sorry for her. I failed hopelessly to make them understand; they had so completely jumped to the conclusion that wanting her head-dress meant wanting her hand, that I found it useless to try and explain. So I bolted into the tent. And when I came out presently with my sketch-book, and tried to take her picture, they concluded that I had only meant that. I never got the head-dress. I had not courage enough for the task. It seemed like asking a lady to hand over her diamonds.

The Uano family were used to my ways, but new-comers like the Yeliseis never got over the feeling against being sketched. Onaska too hated it, and once when catching him asleep on the ground I had succeeded in doing a rapid study of his face, Mekolka woke him up, and told him. He was mad. He jumped up, and ran round and round in a circle, which worked off the evil influence in some way, I suppose. And the sketch of the camp which I give here was done under great difficulties. For always I had a reporter looking over my shoulder, and no sooner did they suspect that I was drawing one or the other than the word went forth, and the subject dived behind the first screen that offered.

'I dabbled about in the river Tinyan to-day, and would have caught a little fish, for I ran him on to the shallows, but dear old Sailor got so excited about it that he mudded all the water at the critical moment.

A CORNER OF THE GOOSE CAMP ON THE PESANKA PLATEAU

Face page 284

I caught one of them later. It is something like our bullhead, and the Samoyeds call it 'har-mùr-gai-ly-i.'

'The cloudberry fruit is now developing. It is green at present, and tightly wrapped up in the calyx.'

Sunday, August 5th.—A wonderfully clear day. All the Samoyeds and Hyland went down to the sea meaning to catch geese. I remained in camp, for I had much flower-pressing and other work to do. They left eight dogs in camp, who never stopped yelling the whole time.

Old Sailor nearly brought down his doom to-day. He has no teeth; but there is something in his English growl—a rolling bass—which answers as well. No matter what dog, however fierce or big, comes up with the intention of eating Sailor, as soon as he, Sailor, stiffens his back and speaks the enemy incontinently draws off. But he is a rash old dog. Absolutely fearless, he will go prowling round the choom, intruding on the others' ground. I always felt that he would do it once too often. And to-day he deliberately went and took a bone from the middle of a group of three dogs. There was a moment's pause of sheer astonishment, and then like a flash the three were on to him. Another second, and Sailor was in the middle of a pack of screaming demi-wolves. Ni-arr-way wasn't there, thank goodness! nor the big brindle; they were at the goosing; but there were quite enough without them. I actually saw one biggish dog catch old Sailor and shake him from side to side. Well, I picked up the tent mallet, and shouting the Samoyeds' word 'wenquoh' (dog), let into them right and left. At the first chance Sailor was out of the *mêlée* and into the tent through the fly in a twinkle. They all but followed him in, and a pretty mess we should then have had; but there was lying there a prop, with which I slipped into them with such effect that they retreated. Well, will you believe me, old Sailor had never a mark, excepting one on the muzzle? Think of the toughness of him; but then he had a coat on him that it would take a long tooth to get through. I wasn't bitten either, and yet I would no more have ventured in cold blood into the middle of that pack of devils than into an active crater.

A cock willow-grouse I shot to-day was changing its plumage to winter dress.

August 6th.—'When the goosing party returned this morning at one o'clock they brought with them only twenty brent, bean, and white-fronted, all told. They had only three boats, and the geese defeated them.

'Uano tells me he saw a Russian sailing boat, and that the ice was much looser.

'Our sugar, jam, and butter, which we have eked out by pinches, is now finished; one doesn't often make two pounds of butter last so long, nor a ham either, for that matter. [And to make a ham last from June 21st to

August 7th was pretty well, you will admit.] It, the ham, is scraggy now.'

August 7th.—We found this afternoon several willow-grouse broods. The chicks of unequal size, the oldest a fortnight old perhaps.

You won't want detailed accounts of our shooting. We never shot but for the pot, or for skins.

The other day I noticed a pair of Arctic skuas hanging round a pool, and they were there again to-day. Hyland walking round found a fine young one in the rushes. We took it home.

The Samoyeds, who go by the sun, have supper earlier every day. I heard such peels of laughter from the choom this evening that I went in, and found them playing with one of my india-rubber paper-bands. Their delight was marvellous. They put it over each other's necks in turn, and snapping it passing it round the circle. But when Mrs. Uano's turn came it suddenly broke, and snapped her on the nose. She looked so startled at the unexpected attack that I laughed, but the others only said 'Ah,' and a sudden great quiet fell on them while they reviewed the wreck.

All this seemed to me so natural. Which of us would not be just as much delighted on seeing such an extreme example of elasticity for the first time?

August 8th.—With a south-west wind and sun to-day

the thermometer in the tent registered 68°, and it felt oppressively hot.

Hyland went down to Scharok with Mekolka, and brought up the news that the ice was gone, and even the gulf looked clear. But that there was no sign of the *Saxon*.

I picked the first flower of *Veronica alpina*.

August 9th.—We had such heavy rain last night that most of my time to-day was taken up with drying our things.

But one most important event must be put on record, because it was the occasion of at least as much interest as a christening in England. I refer to the piercing of the baby's right ear. Katrina did it with a needle, and then put in a metal ring. There was quite a gathering in the choom. The child which was held by Ustynia in the centre of a ring of observers protested very little.

I remarked, when speaking of the admirable way in which Samoyed children are brought up, that they are controlled by kindness. It so happened that on this day we saw for the first and only time a child punished. Little Wanka's father, Verrmyah, wanted the boy, for some reason; but when called he was playing with an old pet reindeer and would not come. His father called in vain three times, and then, without a sign of loss of temper, went up and 'spanked' the boy for disobedience.

Hyland and I were both much struck with this solitary instance of punishment given or needed.

A SAMOYED VERSION OF 'THE PROFLIGATE'

The figures are intended to represent Russians

(*Drawn by Onaska the Samoyed, on Kolguev*)

Between pages 288 *and* 289

CHAPTER II

BACK TO SCHAROK

August 10th.—A day with the swans.

I drove off about mid-day with Onaska and Mekolka. Onaska, by the way, was not only a good scholar, but could draw a little. The native-drawn pictures in this book are by his hand.

He was an ugly but kind-hearted man, who seldom laughed, though some of his pictures betray a real sense of humour. He always seemed to feel the cold severely, and had a way of standing about with his hands tucked under his 'malitsa,' which I was unkind enough to mimic one day, so that it remained a standing joke against him.

We first followed a sleigh trail to the north-east. This trail, covered with grasses and clearly visible even in the sandy soil, was evidently very old, and seemed to point to a time when Kolguev was far more thinly populated than it is to-day.

We passed many old beaches. Some of these, under denuding influences, had been broken down, so that only little conical hills were left. On the top of each of these a fox-trap was set. As in all I visited, the old seal-fat bait from last winter still remained. The adequacy of the summer supply of natural food is well enough shown by this fact.

At last turning to the seaward we came upon the Swan lake. It was a fine big lake, about a mile long and a mile inland from the sea. Round it were flying several lots of grey plover, young and old, and ruffs in winter dress. And away on the other side I could just make out a pair of old swans, and four grey cygnets.

All round the lake it was very marshy, with deep slews, filled with mare's tail (*Hippuris*). But the actual bank of the lake was firm for the space of a sleigh and a team of five; so that we could comfortably stand or uncomfortably lie there, as I did.

Here I found a new and very conspicuous plant—a gigantic fleabane (*Cineraria palustris*) growing in dense bold clumps. Out across the koski was a little ice, looking very loose.

The lake lies north and south. As we drove the sleighs round to the farther bank the swans moved away and swam northwards. We halted and laid our plans.

Onaska was to stay with the deer, I was to hide up in the grasses with my gun, while Mekolka with his bow and arrows was to take the boat which we had brought, and scull after the birds.

So, after much difficulty in wading through a deep and muddy creek, I found a place well screened, and squatted down.

SWAN SHOOTING

Face page 291

I watched Mekolka with much interest. A strong wind was blowing against him, yet he held on his course, sitting with his face to the bows.

Turning a corner he was out of sight, but presently I heard a 'honking,' and peeping out, there was Mekolka, who had turned the swans, and was pulling after them down the lake and gaining every moment.

Now he stopped, and picking up his bow he slipped an arrow at one of the cygnets (they were quite big birds) at about twenty yards, as near as I could guess, and with such effect that the bird beat the water wildly for a moment, and then lay still.

Now he was opposite to me, and the way he had lost in shooting and picking up his bird he was rapidly making up.

He loosed another arrow, and I saw one of the young swans flap along for some ten yards, drop its head, and lie still, while Mekolka rowing up pulled it into the boat. I was astonished that so large and strong a bird should succumb so quickly from an arrow wound.

Then followed an in-and-out chase after the old birds. They led Mekolka a pretty dance, but after a time one of them headed right for the end of the lake, and ran out on to the tundra.

It was a very long way from me, but I could see it running, and after it the two dogs; but well used as they were to catching birds they could make little of this, and though they turned it once or twice it still ran on. I scrambled across the creek, and followed the chase as well as I could over the boggy ground. At last I came up to find the bird had taken refuge in a little lake. Mekolka, who had left his bow in the boat, was waiting for my arrival. He had not thrown his parlka because I had told him not to damage the feathers.

I shot the bird. It was a Bewick swan, in full moult.

I was glad to get these birds by way of proof. For before this I only had the eggs we obtained on the Kriva.

The very bow Mekolka used to-day he had made this morning from a bit of driftwood. I have it now, and his curious little plane.

When we reached camp little Wanka was playing with his reindeer. There is an old reindeer, a capital animal in a team, who is a perfect slave to the child. He will follow him about like a dog. And Wanka spends most of his day bullying this old deer.

First of all he swaggers out, imitating most comically his father's walk. Then with his little di-zha he makes many shots at the old deer (who does not protest), until at last he manages to get it by the horns. Then he harnesses, unharnesses, and generally maltreats it hour after hour. The boy is not strong enough to lift the deer's legs (and that is a very important part of reindeer management), but this matters little, for the old deer well understands the game, and at a touch from Wanka lifts its legs itself. How many times this performance is gone through in the course of a morning it would be hard to say. Yet the good old deer never loses patience.

There is a savage black dog also who came with Verrmyah, Naryàn by name, who bites every one else. But to-day little Wanka was bullying him, harnessing him like a deer, and Naryàn stood there enjoying it.

August 11th.—To-day we moved our camp three versts further up the hill, to try and escape the reindeer fly, which has become an intolerable nuisance.

We pitched on an old camping ground, where for the first time we found a genuine spring of good water at a point where the sand meets the clay. Unfortunately from this point we cannot see the sea; but in the evening I walked off to a hill from which I had a view of the whole coast from Pesanka mountains to Barokoska Nos. I saw no sign of a boat, and only a little ice.

I came upon many packs of willow-grouse cocks, varying in number from fourteen to twenty-three.

Mekolka brought in to-day a red-throated diver in the down. He said that he was after geese in his boat, when the little thing, after diving in front of him, ran out on to *the bank*, when he caught it. A strangely undiver-like proceeding.

Hyland has been given a sweet little puppy which answers to the name of Tyamuk. A most sagacious little creature, though only six weeks old; it is most quaint to see it coming across all alone to our tent from the choom.

Sunday, August 12th.—To-day I went down to Scharok with Uano. We each drove a team of five. As I think I have never properly described a reindeer team I may do so now.

First the names. Reckoning from left to right you have Niasminta, Niasminta Gei-o-hu, Yierrkha, Warrü Gei-o-hu, Warrü.

There is only one driving rein, and that is attached to the head-band of the Niasminta. The off-side of this deer's head-band is then attached to the near side of the girth of the next, and so through the rest of the team.[1] From this it is clear that the Niasminta gives the cue to all the others, and should be the best trained deer. As a matter of fact, he is not only this, but always the finest that can be found.

Driving is a comparatively simple matter. Let us watch a Samoyed off.

Whenever the driver leaves his sleigh he throws the long driving pole down in front of the team, and turning the team itself round till it is facing the back of the sleigh, fastens the Niasminta's head by the rein to the side bar of the sleigh.

This driving pole, which is some ten or twelve feet long, is called by the Samoyeds the 'toorr,' and by the Russians the 'hareh.' It has a button of reindeer horn at the point.

Now the driver unfastens the long rein (the 'moi-ty-

[1] This arrangement is sometimes reversed, so that the Niasminta has more freedom.

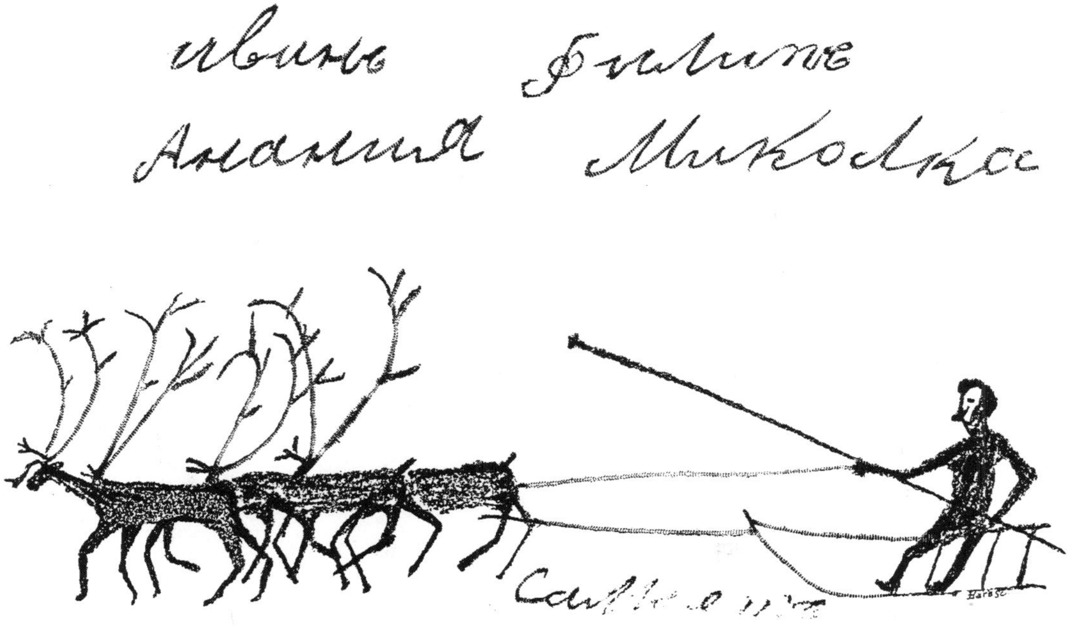

A DRAWING BY ONASKA THE SAMOYED

This picture shows correctly the attitude of a Reindeer driver. Above are in Russian the names Ivan, Philip, Anania, Mekolka. This is a specimen of Mekolka's handwriting after one year's education. (*See* p. 278)

nia,' as he calls it), passes his thumb through a slit in the end, takes a turn or two round his head, gathers up the remainder in coils, picks up his toorr in his left hand, and then, turning the deer straight, runs his eye over the team to see that all is right. For each deer is attached to the front bar of the sleigh by a single long trace running from the collar. The Niasminta on the left or near side, and the Warrü on the off-side, generally ran clear of this trace, one to the left, the other to the right of it, but it passes between the hind legs of the three middle deer of the team.

The following diagram will give more clearly than a description the parts of the harness and the Samoyeds' terms for them.

It is all worth giving for the sake of future travellers, for before this it would seem only the Russian names have been given in the books.

With the exception of the saddle and collar, which are of tanned leather, all this is made of raw seal hide.

So before our Samoyed can start he must see that the 'sach' is in its proper place, and there is usually a good deal of leg-lifting before this is satisfactorily so. Then, leading the Niasminta by the head, he starts the deer at a walk, which after a pace or two becomes a trot. He then stops, letting the rein uncoil from his hand, and as the sleigh flashes past him he throws himself on to it, sitting with both legs to the *left* side. He sits facing his team, the left foot either on the runner or hitched between the uprights, and the right leg either resting on the left knee or on the rim of the sleigh; the rein held in the right hand, twitched up over the 'howlsug' or button as a rest. The long toorr held in the left hand is balanced on the right arm, and during the whole journey is kept waving up and down, now brushed down over the

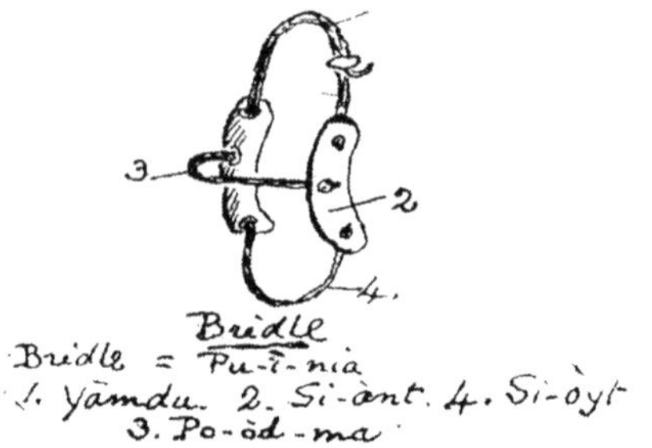

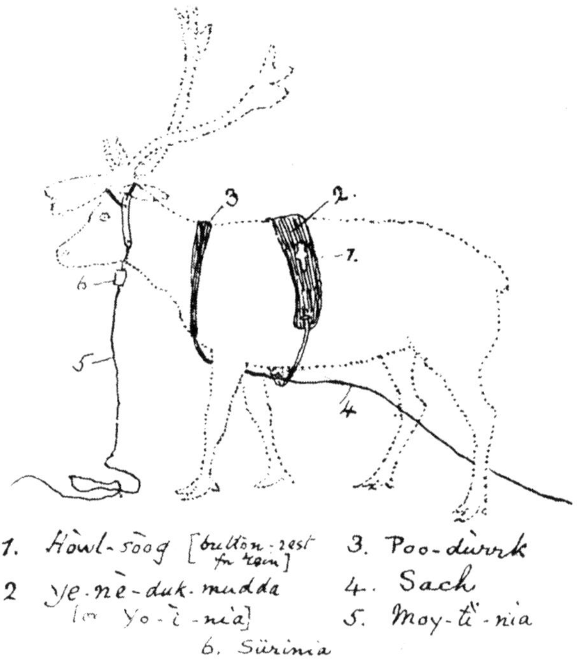

DIAGRAM OF REINDEER HARNESS

rump of one deer, now over that of another. Occasionally, when a deer is very sluggish, it is prodded with the button ('toorr-mahl'), but usually a touch is enough. Also a wild cry resembling 'Hooah' helps to urge them on, and sometimes a hissing sound, such as elderly ladies in a basket-trap address to the pony. If an extreme of speed is required then the driver raises both arms, a signal the deer know well.

This generally puts them to a gallop. Are they to go to the left—then the rein is simply pulled; if to the right,

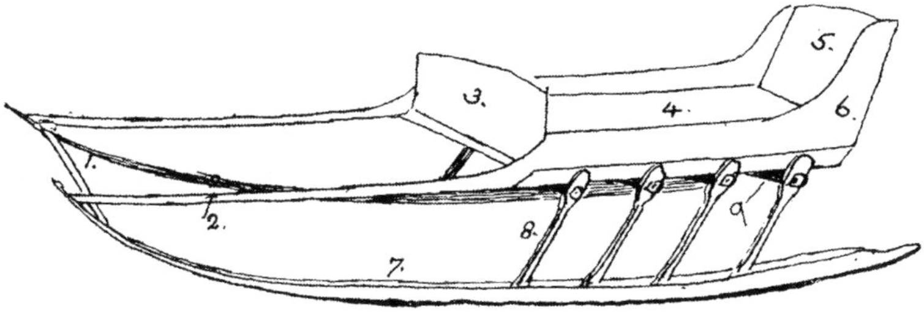

DIAGRAM OF SLEIGH

1. Sid-der-nla-dit-sa.	4. Han làta.	7. Han gaz.
2. Hannin.	5. Poy-nld-du-òn-durr.	8. Sit'n-dl-ek-nya.
3. Tchòy-dùrr.	6. Hanwàno.	9. Ni-àrrt.

the rein is drawn sharply across Niasminta's quarters, and the whole team turns that way.

It is easy enough to do all this; but one thing I could not learn. It is—how to make the reindeer slacken speed. I never could understand the Samoyeds' own explanation; I never could determine how they did it themselves. So any reader who has learnt this secret (and doubtless many have who have travelled with the Lapps) can do me a kind turn by making the point clear.

A Samoyed has not room on his little sleigh for very much besides himself, but three things he never moves without—his rifle, his axe, and his sovik.

His rifle is bound to the sleigh. Every Samoyed has an old flint-lock rifle besides his bow and arrow. The bullets he himself makes in an ingeniously-made mould formed of stone set in a wooden pair of scissors. His axe is of the Russian pattern, with a nail extractor on it. The sovik I have described. He sits on it, and wears it at night when sleeping on the ground.

So Uano and I went down to Scharok. I was driving by far the finer team, none the less Uano, as the better Jehu, more than held his own. He could get more work out of his deer, by some silent skill I could not comprehend, than I out of mine, with all my prodding and noise.

Arrived at Scharok I was delighted to find the ice entirely disappeared. I fixed high up on a pole the two yards of red flannel I brought for a faja or kummerbund, and there it waved, a noble flag. Also I wrote a second note to Powys, and nailed it on the flag-staff, where I thought it might be better seen.

Visiting the site of our camp I was disagreeably surprised to find that a fox had scratched a passage into the cache and dragged out several of the brent. I did not mind that, but was sorry only that he had let in dorndaftsa, the blue-bottle fly. So I set two traps to take him there.

A SAMOYED SHOOTING REINDEER

The man is screened behind a lump of turf in which reindeer horns are fixed as a decoy

(*Drawn by Onaska, on Kolguev*)

SAMOYED SHOOTING FOXES IN WINTER

(*Drawn by Onaska, on Kolguev*)

Face page 298

As we were passing Solnoi Toh on our way up, I picked for the first time grass of Parnassus in full flower.

We made many halts on the journey, both coming and going, and once I tried to stalk a fine cock snowy owl; but creep as I might, before I was within gun-shot off he grandly skimmed.

And while Uano and I were smoking and taking snuff the old man told me many interesting things, some of which I noted down.

I asked him about the white reindeer, how they came to be white; for I told him wild reindeer were not white. But old Uano was never very good at reasons.

They were white, he said, because their mothers were white. 'Wazanka white, Teliurnok[1] white, yes, yes,' was how he put it, and I could not get him further back than this. So then I asked him which were best. 'Both best,' he said; 'yes, yes, white legs strong, black legs strong, yes, yes. Both best.' He always called grey black. But pantomime was Uano's *forte*. He had been saying to me over and over again, 'Black eyes no good, no good; white eyes good.' And I couldn't imagine what on earth he meant. So he sat on the ground studying and studying a long time to himself, evidently quite distressed at his failure to make me comprehend. All of a sudden off he ran on hands and feet, scuttling over

[1] 'Teliurnok' is of course Russian for a calf. 'Wazanka' is a corruption of the Russian *Vashinka*. It is always used by the Samoyeds for a reindeer hind.

the tundra like some old badger. 'Troobka, troobka' (telescope), he shouted. 'Uano black, yes, Uano quite black, yes, yes, no good.'

Even then I could not form an idea as to his meaning. With that he comes up, impatiently seizes my glass and says, 'Now Ahnglia reindeer, yes, yes.'

So off I went in turn at this strange game; I was to be the reindeer this time. Then he shouts, 'Reindeer black, no good; eyes no good, many tears.[1] Now reindeer white, good eyes, very good, yes, yes, laughter, Verrmyah's adliurs. Yes, yes, reindeer come quick.'

And then I saw it all, as of course you will by this time. Translated it only means, 'White is a much better colour to see than black. So white deer are an advantage when Verrmyah goes out to find the deer.' And this was quite true, for a grey reindeer is very hard to see against the grey peat.

Many other things Uano told me that day; where the fish were, and how the salmon trout only ran up two rivers, the Gusina and the Kriva, to wit. He told me that the Kriva rose in a big lake. The lake ('toh') he called it. He told me of high mountains,[2] so steep that reindeer could not pull the sleighs up their sides. Many things he told me, but one he would not tell. He would not tell me where the brent geese bred. This was a

[1] This word 'yarrn-ga' is literally *tears*. It is used to express the watering of the eyes from the smoke of a wood fire, but also they use it to express any defect of sight. So 'pisinga'—laughter—is used for happiness or success.

[2] Sowandeyi and Siecherhur.

SAMOYEDS KILLING WALRUS

The black dots on the right are heads of seals at which a man, lying on the ice, shoots

(*Drawn by Onaska, on Kolguev*)

A SAMOYED TAKING A DEER WITH THE DI-ZHA

(*Drawn by Onaska, on Kolguev*)

Face page 300

point of honour with the Samoyeds, and I could not get him to lend me a team of deer so that I might go right away. All the deer were wanted for the goose catching.

August 13*th*.—At half-past one in the morning we came again in sight of the encampment, and we made a race of it as we drew near. My fine team seemed to be doing all they knew, but at any rate Uano's team did more, and I was a bad second at the post.

Then Uano invited me to drink tea—my tea—in his choom.

I noticed that we sat down with rather more ceremony than usual, and Uano, who generally sits on the other side of the choom, made me take a place on the tub by him. And then I found the reason. As I had given them real tea, Mrs. Uano was giving us real tea-cups. Out they came, two Norwegian and three Russian cups and saucers. They had been a present from the wrecked Norwegians.

Then Uano told me how every year he bought a hundred pud (a pud = 36 lbs.) of flour from the Russians, and that for this he gave two roubles per pud, and that his Russian's name was Alexander Samarokoff, merchant from Okshin on the Petchora; that this Russian and another came east in his boat every year, and that had it not been for the ice they would have been here ten days ago. And he brought me his wooden calendar on which the dates were ticked off.

'They will never come now,' he said; 'very much ice, ship lost, yes, yes.'

I got to bed at 3.30, and slept right away till 9 o'clock, an unusual feat for me.

And it does not seem that I did much of interest this day. I caught a great quantity of *ephemeridæ* by a pool, and I found the crop of a young willow-grouse full of the buds or spores of *saxifraga cernua.* Also, I opened my beetle box, and found to my sorrow that a beetle I had believed defunct had recovered and had eaten all the rest.

August 14*th.*—Hyland still suffering from headache. I went round a small chain of lakes by myself, but saw little of importance.

Little Wanka was very comical—the whole day he went stumping around in his father's long boots, saying, 'Ahnglia, Ahnglia,' every time he passed our tent door He meant he was like me, with my big boots.

Two notes from my diary are—

'A barrel full of salted geese contains 300, and sells for ten roubles. The value of the barrel is two roubles.'

'The two largest calves here are 3½ months old, *i.e.* they were born the first week in May. When ten months old they are entered to sleigh work, the five going straight into training together without any older beast to steady them; Verrmyah said this was great fun, "much laughter."'

August 15*th*.—I fear if I make more digressions this will run into a very big book. So now for diary pure and simple.

'Wind SE., light till evening, then strong. Heavy hot shower at 2 P.M.; thermometer, noon 67°, midnight 52°, bar. 29·7 in. I went out early for a walk by the sea; it is a lovely morning, and I exactly determined our position, having good fixed points in Pesanka range, Barakoska Nos, and Swan Lake. Collected some snowy owl pellets from a mound, on which one, a male, was sitting. There are, I believe, no mammals' bones among them, excepting in one, very old and bleached. Here I found a few bones, probably of a lemming. The deer have eaten the place clean, and are now breaking away so far that Uano said we must move to-day. At 5.15 therefore we left, and at 6.45 we reached our present camp. Very slowly we came, winding in and out of the hills, till we rose 150 feet to an excellent little hill. South-east of this is a rise in the ground, from which you can see the sea all round, and even the huts at Scharok. I went up there with my glass, but could see no sign of a boat. They killed a cow (wazanka), and I bought three funt (about 3 lbs. 14½ oz. go to the funt or foont) of fat at 10 kopek per funt.

'There was a very curious effect this evening, and a beautiful picture as the moon rose; for the moon was huge and red and oval, all its mountains and hollows sharply defined. In the foreground the tundra, with the mist filling in all the hollows, and the reindeer just visible as they walked off feeding to the skyline, completed a picture which I stood and admired long. My wretched sketch gives no idea of it. Hyland shot a titlark, one of a flock.

'Old Ustynia has been very hard at work making a bridal dress or panitsa for young Ustynia, for she is to be married to On Tipa's eldest son. The dress was constructed of many small bits of reindeer skin, stitched together with sinew. The sinew is saved from the back of the deer, hung up to dry; it is afterwards torn in strips, and each strand before being used is rolled between the hands as cobblers at home roll their thread. The women use sometimes a bone thimble on the thumb. And they stitch outwards, that is to say, away from themselves, as sailmakers do. They stitch very nearly as quickly as Englishwomen, in spite of their contrary way. Their needles, which they get from the Russians, they keep with various odds and ends in prettily ornamented skin bags.

'Old Ustynia begged two pieces of wire to-day for knitting needles, and her delight was tremendous when I also gave her a pair of woollen stockings.'

MEKOLKA'S CAPTIVE—'TCHUÒRNIA,' THE RED FOX OF KOLGUEV

Face page 305

August 16th.—This was our record day for heat. The diary runs :—

'Wind S., very light—thermometer, 7 A.M., 70° in sun ; noon, 76° in shade, 86° in sun ; midnight, 54°, bar. 29·9 in. Clear all night, which is most unusual. The hottest day we have had. I went into flannels. The dogs dead beat, lying about just as they did last Wednesday, and yet the heat would not compare with that of a hot English summer day. Even the Samoyeds felt it, and opened flaps in their choom. They lit their fire outside for the first time. The girls came out to-day in red skirts and belts ; the men also. They wear the shirt-tail outside in the Russian peasant fashion. (Baby was hawked about all day without its cradle or its clothing.)

'The bulls' horns are now beginning to lose the velvet, but the steers retain theirs.[1]

'In spite of the heat we had no musquitoes to-day, no doubt because of our high position. But the fly was very bad indeed.

'While Mekolka was out adliurgo to-day his dog caught a young red fox. It was little hurt, and he brought it back. He had tied its legs together, and muzzled it also with a leather thong. I was sketching it as it lay on the ground, when suddenly it slipped the thong from its feet, and was off down the snow ravine at the back. Instantly every man, woman, child, and dog

[1] The majority of the gelded deer never lose the velvet.

was after it. Mrs. Uano stood near the bottom of the gully, where she had been getting snow for water. Hearing the hubbub, and seeing the fox coming, she dropped her pot and stretched out her arms to receive the fox (who ran straight at her), made a grab at it, missed her footing, and went such a sprawler down the hill that all the field screamed with laughter. In a second the dogs had caught the fox, and there was a fine muddle. I stood at the top and laughed, for it really was comical. The bottom where they caught the fox was all mud and snow—a fine mess. You could have covered the whole thing with a sheet, and what with the mud and the snow, the dogs and the fox, the question was, which was Mrs. Uano? However, she was very good-tempered about it, poor old thing! The unfortunate fox was badly pinched, and does not look like living now.

'Verrmyah has been working hard all day making me a model sleigh.

'Uano and Ustynia went off to call on On Tipa, and returned in the evening, Ustynia looking very grand in her new panitsa, sitting on her father's sleigh. She brought back with her a sack filled with curious odds and ends of mouldy bread and finery.

'I had a long walk, taking the hill twice to look for *Saxon*, but the mirage was hopeless; even the Scharok huts seemed like some immense fortification.

'It has been such a lovely evening—a blood-red moon opposite the setting sun. The moon again very curiously refracted. For, rising as it seems in quite a clear sky, its slope is oval (depressed above and below), and its upper and lower edges all jagged. As it rises this condition gradually changes, till you have a circular moon with clear-cut rim. Seen through my little telescope it really seems that you could just step down into its valleys—they are so sharp. Behind our tent the gully's sides are covered with snow, much of which is pink with *protococcus nivalis*.

'A merlin flew close by the tent this evening, turning its head, and looking in at the door as it passed.

'They trimmed some calves' ears to-day (*i.e.* cut off the tips for marking purposes), and the women forthwith skinned and ate the bits.

'A yellow-ringed humble-bee stunned itself by flying against the tent rope.'

Ustynia, I should explain, did not drive her own sleigh to-day, but sat behind her father on his because she went to her betrothal. Young On Tipa, whom she was to marry, was not a bad young fellow, but consumedly conceited because, as his father was the richest man in Kolguev, he had been the catch of the island.

The curious part of this Samoyed courtship was that we never saw the man and his girl together. I never saw On Tipa the younger inside Uano's choom, and I cannot remember that I ever knew him and Ustynia to exchange three words with one another.

August 17th.—Now here is an odd fact. The musquitoes were about to-day, yet they did not worry me at all. But Verrmyah, who was working at my sleigh, was bitten and bothered very much. A native might have been expected to be musquito-proof.

On Tipa was to choose three young bulls from our herd as part of his son's settlement. One would have supposed that he would have just jumped on to his sleigh and come across to look over the herd, as a farmer would at home. Not a bit of it. This was not consistent with his dignity on such an occasion. So Verrmyah and Mekolka had to round up the whole herd of five hundred beasts more or less, and drive them twenty versts away to On Tipa on the head of the Veliki river. This they did to-day, and brought them back.

Hyland took my glass up to the hill, and coming back, reported a sail at Scharok! This Mekolka also had seen in the morning. Uano was very cunning about it. He didn't know that we were aware of it, and so he had made up his mind that he would slip off by himself and see his Russians. I think that he chiefly wanted the pleasure of telling them all about us, and of talking us over with them before we met. So I did not even know he had started until he was two miles off. Then I went for Verrmyah, who said, 'My adliurs Ahnglia's adliurs. Ahnglia go quick, yes, yes.' But I had no intention of pounding all the way to Scharok for nothing. For of course we knew the sail was not the *Saxon's.*

So I wrote a third note to be left in the chapel, and sent Verrmyah galloping off with this after his father.

This was a memorable day because of a great thunderstorm. At one time it was pouring rain and flashing lightning on every side. Only where we were was a single clear spot.

Our poor Samoyeds were very frightened by thunderstorms. They were good Christians while it lasted. At every extra big flash or peal they crossed themselves religiously. Mrs. Uano brought out a little ikon (a picture image) on a stick, which she stuck in the ground. When first we came we saw a good deal of this little ikon. (I think it was St. Vasili of Solovetsk.) For we were then supposed to have come from the Governor of Archangel. But in the familiarity of these later days poor little Vasili had been kept in the bag, and now he was only brought out for thunderstorms.

To-day Mekolka brought me in wonder a common earthworm—just *lumbricus terrestris* to all appearance—which he had found lying on the ground. But I had no spirit for its preservation. It interested me much. When one considers that the ground for seven months out of twelve is frozen solid, it seems strange that a creature so highly specialised as the earth-worm, a creature too, whose development is direct, should be able to survive. But Mekolka had never seen a worm before, and this one, I suspect, had been brought over in the crop of a gull.

August 18th.—'This *has* been a day.'

I was sitting at the tent-door at five this evening, making a sketch of pi-liuh, the reindeer fly, and his larva, when suddenly away on the tundra rose a most extraordinary yelling. It gathered force as it came on, and lo! it was Uano come back from Scharok.

He was 'half-seas over' now, and had evidently been right across the ocean.

'My Russmann come,' he screamed at the top of his voice, and again 'My Russmann,' and a hundred times the same burden, till at last he pulled up and tumbled off the sleigh, quite helpless.

They helped him into the choom, where he soon recovered, and volubly gave an account of his visit. I could hear it all going on, but had not the spirit to go in; it was so depressing to see our kind old friend come to this pass. Now and then he would shout for me, but I would not go. At last he came stumbling across to my tent, and plopped down in the entrance.

'My Russmann,' he says between crying and laughing. 'Uano's Russman come Scharok! My Russman very good! yes, yes. My Russman say, "I wish Hylum and Ahnglia, no, two Ahnglia—here—tea, sugar, tobacco, yes, I give all." My good Russman! yes, yes.'

'Two Ahnglia' (two England) showed that the Russian had tried to explain to him who we were.

But he was such a nuisance that I made his wife come down and drag him off.

Poor Hyland, who was out, had a second performance on his return all to himself. I am afraid he took it much more good-naturedly than I.

Later on came little Wanka to my tent, bringing the first written invitation I had received on Kolguev. It was evidently Mekolka's effort, and it ran thus:

'Ivan zalaht choomoo,' which composition meant, 'Ivan (Uano) desires your company in his choom.'

Hyland and I went. It was a tea-party. They put me in the place of honour on Uano's right hand. We had real tea in real tea-cups, sugar, mouldy white bread, little rings of bread containing carraway seed (the Russians call them kallatch) and vodki. It was a great time for them; though the vodki caused a good many sad things to happen that evening. But I noticed Mekolka was not there. I myself had taken one glass just to try the vodki, and please these poor people, and Hyland told me afterwards that Verrmyah had said I was just like Mekolka—for Mekolka would never have more than one glass. This we found to be true, and it was very remarkable in a semi-savage. But then Mekolka was quite a remarkable chap.

Truth to tell, this vodki was so abominable that had either Hyland or myself been fonder of spirits than we were, I think it would have choked us off. But Hyland was one of the most temperate men I have ever met. He often sighed for a gallon of beer, poor fellow! but sigh you might in Kolguev.

I left very soon on some pretence, for I longed to escape from the pitiable into a purer air. So I walked north a long way till I found a little gully, where I stayed.

One side was all snow, the other grassy and ablaze with flowers.

Here I found a double red raspberry, and a beautiful tall veronica (*V. longifolia*). And there was a ringed-plover evidently with young.

'A dense fog came on,' says my diary, 'while I was in a perfect labyrinth of gullies. But I think I must be getting the homing instinct better developed, for I walked straight back, though I could not see the choom till within a few yards.'

Sunday, August 19*th.*—Our friends spent to-day in recovering from yesterday.

We caught a wazanka with the di-zha for the Russian's food. I got old Ustynia to milk it for me, and thought the milk exceedingly good. It was like cream. But the Samoyeds kept saying 'no, no, not good. Very bad, yes, yes.' It is strange that these people never use reindeer milk, for the Lapps, we are told, drink much of it.

Mekolka comes in every day for a lesson in English. The English he writes phonetically in Russian characters, and remembers it quite well. At present we are hard at work on the numerals, and it is very funny to hear his attempts at pronunciation. The only one which really

VERONICA LONGIFOLIA

defeats him is six. He goes over this again and again, but always with him it is 'sikkus.'

August 20th.—At last our stay with our simple friends was over. We were to share their society and that of the Russians from to-day.

We went in a fine procession down to Scharok. There were two sleighs laden with our things. Katrina was there, and baby, and Mrs. Uano in a brand-new scarlet cotton Russian dress. The effect of this was striking, for as she had pulled it on over her old skin panitsa, not a button of them all would meet. Really she did look very funny—something like a toad in a cactus flower.

Verrmyah's warrü was bogged on the way, but we managed to get it out.

We raced into Scharok. I was driving my old team of five, and beat them all but Verrmyah. I never could quite catch him.

The appearance of the huts was already changed. Fires were burning, and odds and ends of tackle lying all about. From the mast-head floated the red flannel flag.

The principal person who came out to meet us was the trader himself, Alexander Samarokoff. I admit I was agreeably surprised. I had expected to see some low ruffian trader, and instead here we had a handsome, bronzed, black-bearded man, with clear, keen grey eyes, and a manner absolutely self-possessed.

I said little to him at first, for I intended that he should have the idea that we were entirely independent of him. Poor old Uano was wofully disappointed. He had evidently set his heart on showing us off. But caution is wise until one knows one's ground. And it seemed to me that, as it might possibly come to a question of *compelling* him to take us back, it was better to keep the high hand till I knew my man. So I said, 'How d'ye do?' as though we often met. Then I introduced Hyland. Hyland's fine big figure and fresh English face evidently took his fancy very much.

But I immediately set to work to move our belongings down to the site of our old camp.

Before going I asked but one question, 'How long do you remain on Kolguev?'

And Alexander Samarokoff answered, 'One month.'

PART VI

TELLS of the Holy Hills and of household gods—Of Samoyed labour and of traders' ways—Of the lamp of St. Nicholas and the bringer of storms—We leave Kolguev, and after much painful effort reach the mainland, and finally home

CHAPTER I

TO THE HOLY HILLS

August 21st.—Now it is time to explain something of these Russian merchants with whom we were to live.

Years ago—fifty years ago—in the days, perhaps, when the priest had visited Kolguev, there had been quite a considerable traffic between the up-country traders, men of Mezen, Indiga and the Petchora, and the Samoyeds of Kolguev. The Samoyeds had lived there always, I believe. I could never learn of a time in their traditions when they were not dwellers on the island.

They were very useful to the Russians traders. Not only did they supply them with skins of polar bear, walrus, and seal, the feathers and down of ducks and geese, but also they were pasturers of the Russians' reindeer. They entirely owned a proportion of the deer, but also they tended many herds which belonged to the Russians solely, or in which they had a half interest. In those days there were, according to Alexander Samarokoff, no fewer than 25,000 reindeer on the island, of which 10,000 were owned by one man, Alexis, a Russian.

Then there fell the first of those terrible reindeer

plagues which were to bring such havoc into all those parts. The reindeer succumbed by thousands, and the trade with Kolguev was paralysed and died. To-day no Mezen or Indiga traders can tell you anything of Kolguev. They know not whether Samoyeds are there or not, or if there, whether they are only visitors, going and returning in the summer months. They know nothing. That page in their history is clean wiped out.

But it chanced that two brothers Samarokoff, merchants of Okshin, a little town or gubernia on the Petchora, continued going. Why should they not? There was less competition now; at least they could get from the Samoyeds all the produce of their hunting in exchange for their cotton goods, or snuff, or tea.

They were dead long since, but their sons, Alexis and his cousins Alexander and Ivan (now dead) kept the practice alive; and Alexander it was who was here now. Alexander came first to the island with his father as a child of ten years old thirty-five years ago. Not for a single year since that time has he missed the voyage.

On this evening we received from Samarokoff an invitation to tea. It was brought to us by one Yakoff Popoff, his servant and factotum. This Yakoff was a quizzical sort of character. He had a prickly snub nose, small twinkling grey eyes, and long straight reddish-yellow hair. He had served as a soldier, and was extremely proud of his personal strength, though he was not very strong. He had a great deal of the oracle about him because he had been to St. Petersburg.

Alexander received us in his hut, where he fed us on raw salmon, raw flounder (which they had taken with their nets from the creek), some milk and vodki; and afterwards with flounder cooked, cooked reindeer, rum and tea.

We stayed long with him, and he told us many things. He said that he feared his cousin Alexis was lost in the ice, for he had left the Petchora in the beginning of June, and had not been heard of since.

August 21st to *24th.*—Well, all these days passed without any event of moment. On the first we picked a tin-full of cloudberries, not ripe, but nearly so. The next day Mekolka came down, and brought with him my little sleigh and a complete set of miniature harness. On the following day the blue fox came back for another look at the geese, and finding the spot occupied sheered off, and came cantering jauntily past me within ten yards.

Aug. 25th.—On this day it was arranged that I should go and visit the high mountains. Alexander, besides having a part share in Uano's deer, had a herd of some two hundred or so away in the hills near the rising of the Gobista river. These deer were under the charge of a second Marrk (not our old goosing friend) whom we had not yet seen. And in the course of the afternoon he turned up with reindeer—with two teams.

Hyland, who was ill with a chill, was not able to get

up all day. I sometimes had an idea that, with a little resolution, he might have shaken off his complaints. I had told Alexander that he was not well—had a chill. I did not know what chill might be in Russian, but explained that the cold and wet had affected him.

'Ah,' says Alexander, 'I know, "pros-to-deelsa," yes, yes, "pros-to-deelsa."' (The Russians often used this word; it means, I take it, 'with a cold.') So down I go to Hyland, and coming into the tent with a very serious face, I say, 'Hyland, you are very bad. There is no doubt about it. What you have got is nothing less than *Pros-to-deelsa*!'

'Pros—what, sir?' says Hyland, in much concern.

'Pros-to-deel-sa,' I repeat, 'a Russian complaint.'

'Good gracious!' says poor Hyland, looking wofully blue, 'you don't really think so. How ever could I have taken that?'

'Well,' say I, 'I don't know how you took it, but you've got it. Now I'm going off for a day or two, and there will be no one to nurse you, so you must shake yourself together. What you want is a good sharp walk every day. That will put you right.'

So I left. We went S.W., crossing the Barakova, through a marshy country, and gradually rising some hills. Here we were overtaken by a thunderstorm, and old Marrk often stopped to take in Dutch courage from a bottle of vodki. I tried to give him a little lecture on temperance. 'Marrk,' said I, 'vodki not good.'

'No,' answers Marrk; 'Russmann's vodki *not* good, *not* good,' and with that he swigs again. And thereafter, as often as he might drink the vodki, he would shake his head and say profoundly, 'Russmann's vodki *not* good, no, no, *not* good.'

Mine was such a vile team that I gave up driving, made Marrk take my toorr and hitch my sleigh to his. My warrü was a heavy, hornless, sulky bull. We tied him afterwards to the back of my sleigh, where he did his best to choke himself in his efforts to hang back. He moved with a heavy rolling action, very distinct from the pace of the younger animals.

So we kept rising and rising until at last we came upon a single yierserk by a little lake. Inside this I found the old woman had arranged for me a fine white reindeer skin, and insisted that they should sleep in the open. It had rained so hard during the last two hours that everything was flooded. However, when I crept out to see how the old people fared, I found them both snoring, completely smothered in skins.

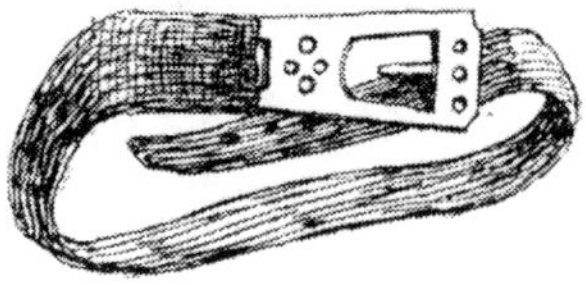

MARRK'S BELT; SHOWING INGENIOUS BUCKLE OF BONE

Sunday, August 26th.—The morning broke so clear and fine that I turned out early and walked all round, taking stock of the country, which was all of the same formation; stony ridges with peat in the hollows and levels, and a river winding through. It was curious to find one's-self alone with these two old people, away on so desolate a spot.

When I returned from my walk they were bringing in the reindeer. Mrs. Marrk was really a remarkable old woman. She and her old man managed a herd of 300 reindeer every bit as well as we could have done with all our hands in Pesanka camp. The old lady was a miracle. She did all the business of the adliurs, galloping round with the dogs, and jumping on and off her sleigh like a young thing. Here is a picture of her taken on the spot, as she sat in Alexander's hut with a pooziri bottle in her hand. These pooziri bottles are glass globes, which the Scandinavian fishermen use as floats for their nets. They are picked up on the Kolguev coast, and then serve the purpose of vodki bottles.

Marrk's Niasminta cut up rusty just after we had left, and the old man was all but hurled over a precipice, sleigh and all.

We crossed the Pugrinoy river, 21 versts from the gulf and 3 versts from its source. Soon after this we came to Honorohur mountain, which the Russians call Lodka (the boat), for that is the appearance it presents to them when seen from the Pugrinoy mouth, where their boats sometimes lie.

I walked to the top; it is about 100 feet high, and had a fine view over the surrounding country. Away to the west were the Gobista mountains, and shining in the sunlight to the south was the great lake Promoinœ, looking as if only a mile or two away.

Here I found a lovely delphinium (*D. elatum*): identical, it seems, with the old-fashioned blue one of our gardens. Some were dark violet, some were blue; some with more, some with less white in the eye. So that which I had always supposed a gardener's creation is really a natural feature.

An hour later we had reached the mountain Sowandeyi. Here was a really remarkable view down into a water-formed crater. But no rock was visible, only landslips of sand. A peregrine falcon was flying round here and crying.

We left at noon, going south-west, and dropped down and down till we reached the Gobista river running west.

Three versts from this we reached the Kriva lake. I looked at this with much interest, as the source of the Kriva river, the stream by which Powys and I had first landed on June 16th. But the lake was not beautiful in itself. It is too square in shape. I calculated, as we drove round it, that it was about 1 mile wide and some mile and a half long.

We found here drying grounds for fish. And when we were about half-way round we came upon the eldest of the Bulchikoff brothers, come to fetch away a barrel of salted fish. He said they were quite good; but they smelled horribly. These men will not take the trouble to salt their fish before it gets bad. The fish were sik, a kind of bream, and besides this, they told me, the lake held only goletz, which ran up into it from the sea. Goletz is a species of *salmo* or of *coregonus*. I am not quite sure which.

I inquired of Bulchikoff how deep the Kriva lake might be. He said, 'My toorr—more big, yes, yes,'—it was deeper than his toorr. But I wanted to know exactly. At last he laid his toorr down, and walked away from it two or three paces, and said, 'Two toorr—no, my big toorr, my little toorr, yes so, yes, yes.' His was a twelve-foot toorr, and the small toor would be a ten-foot one. This would make the depth about twenty-two feet.

So we left, and crossing the head of the river where it leaves the lake, wound round the lake itself. I saw a glaucous gull hammering at something by the side. It rose as we approached, and flying off left behind it a fine sik dead on the stones. It had its guts torn out, but was quite fresh and clean, weighing, I should say, two and a half pounds. Marrk said it was very big for that lake. I was much struck by a curious adaptation of the upper lip. This is modified into a horny-looking beak, which appears to be extensible, and is evidently designed for rooting among the stones. It had several pebbles in its gullet, two of them larger than hazel-nuts.

So, winding round the lake, we took again to the mountains, crossed the Gobista twice, and headed for Siecherhur, one of the four sacred spots on Kolguev. The others are Bolvana mountain, the Pesanka mountains, and Pesanka lake.

This mountain, the highest in Kolguev, is a conical hill, situate some 150 feet above the sea, and rising suddenly to about 100 feet. Its sides are steep and sloping, and it is formed as regularly as an ancient camp. On the top are two very old wooden bolvans,[1] too large to carry away, and many remains of reindeer.

You heard in the saga how the dying man bids his son carry the little sleigh and the spoon to the Holy Hill, where is the sacrificial place of Nûm. I believe that this Nûm is the ancient god of all the circumpolar races; that he is common to the Eskimo, the Chuckches, the Ostiaks, and the Samoyeds. Originally,[2] when a Samoyed died, he was laid out straight on the tundra, and some of his

[1] Bolvans are wooden images of the god Nûm. A face is carved on a bit of wood, and cuts representing ribs are made in the side. This is all. The Samoyed name for a bolvan is 'shyadēy.'

[2] So Uano told me when explaining the saga, and the traders afterwards confirmed the statement.

personal belongings were taken to the sacred spots as an earnest that he was really dead. Old Marrk was wofully alarmed that I, an unbeliever, should visit this place. I think he feared that the wrath of Nûm would fall upon me, and perhaps on him, as he had brought me. So, reining up the deer at the foot of the hill, he flung himself flat on the ground and buried his face in the moss. As I climbed slowly up the hill I often turned and looked back, but always the old man was in the same position.

When I reached the top, which is flat and round, and some fifty yards across, I first examined very carefully the two old bolvans. Then I turned to the view. It was very striking.

On the south rose Somandeyi and Honorobur; to the south-west lay the Kriva lake. At the mountain's foot the tiny Gobista stream went trickling away among the grasses. Immediately below me were the two reindeer teams, and old Marrk lying there as still as an old grey stone, face downwards in the moss. He did not stir till I was again by his side, and I had to poke him up. Then he never turned his head, but, with face averted from the sacred hill, he mounted and drove off.

When we returned to our little camping ground young Kallina had arrived with his wife and his baby. They pitched their little yierserk alongside ours.

We had a lot of trouble on this evening; for several of the bulls had their horns clean and were fighting and bellowing. Marrk had foolishly left two teams unwatched. In each of these was a bull with clean horns, and suddenly they began fighting. And before we could intervene there was a fine muddle. The two teams were all entangled up in a most perplexing manner. This would not do. So we had to throw three of the bulls, and saw off their horns.

Marrk and Kallina threw the bulls, and by a very bad device, as I thought. First of all the fore-legs were tied together, and then Kallina, taking the thong, pulled the fore-legs forward, while old Marrk hauled away at the near hind-leg. The bull did fall at last, but not before he had scrambled about all over the place. I explained how much better was our plan of bringing all four feet together, but they didn't see it.

Such a saw as it was too—home-made and villainous. I really think Kallina must have been half-an-hour in sawing through two horns, while I held the beast's head down.

The Marrks' evening meal was a very fair sample of our feeding under the catering of the Russians. We had five courses: 1. Tea, cloudberries and kallatch. 2. Raw, bad salt-fish. 3. Roast ribs of reindeer (very good). 4. Boiled, bad salt-fish. 5. Boiled reindeer and black bread.

August 27th.—I had a lovely walk very early up the Pugrinoy. The morning was one continuous sunny sparkle. I collected many new flowers—a large ranunculus, a very pretty pink (*D. superbus*) and a gentian in seed (*G. verna*).

At noon we left and and drove back to Scharok, where I was sorry to find Hyland still ill with pains in his back.

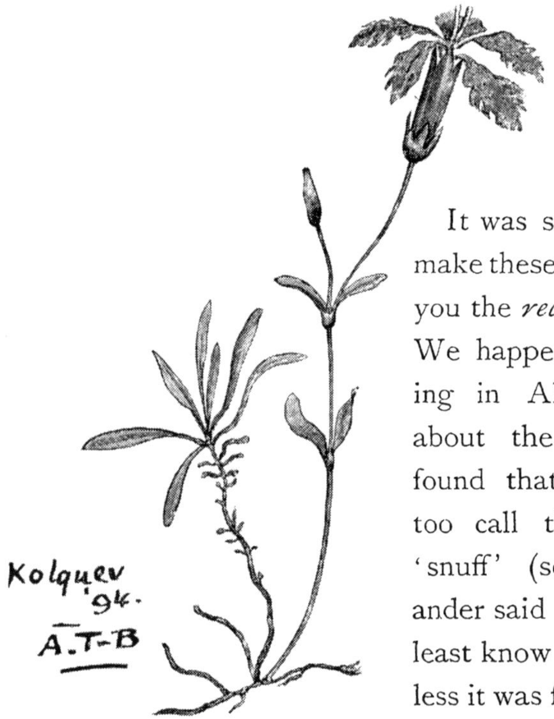

THE SPLENDID PINK (*Dianthus superbus*)

It was so very hard to make these Samoyeds give you the *reasons* for things. We happened to be talking in Alexander's isba about the birds, and I found that the Russians too call the snowy owl 'snuff' (sowah). Alexander said he didn't in the least know why it was, unless it was from some noise they made. I, who knew, of course, that it would be because of the snowy's *hissing* noise, tried hard to get the Samoyeds to explain why they also called it 'snuff,' or 'sneezing' (heinibchur in the Samoyed, as I have said before). I did not succeed, though Uano was evidently quite satisfied that he had given the only explanation when he answered that 'his name was Uano, and mine was "Ahnglia," and a fox's name was Tchuornia, and a bird's Tierrtekoh, and this Heinibchur; and if everything had the same name, how *could* any one tell whether you were a man or a bird!'

August 28th.—This day, too, is worth recording.

It was very wild and rainy, and my second thermometer was broken by swinging against the tent. So good-bye to all my records. Also it took me three-quarters of an hour to light the fire for breakfast, and when it was lit, fine sand blew into the pot and made things gritty.

Alexander, after having lent me a team for the Pugrinoy, determined to go too. So I made young On Tipa, who was there, drive me, for I had found it hard to use my surveying instruments and write my records when I had to look after a team.

He proved himself not only conceited, but what was worse, a conceited fool. For, although I warned him that the tide was up, he drove clean into the river, and not only soaked himself but me, my gun, and everything I had.

After seeing the country I felt that dirty Shabla had been kind in his warning about the creeks. For there were many places we could really not have crossed on foot.

We crossed the Pugrinoy about four versts from its

mouth, and then moved down to the southern extreme. Here we found three old isba and three ombara. One of the isba had a fine brick oven. The bricks for this, Alexander told me, had been brought the whole way from the Petchora by his brother who was dead. Alexander is now guardian to the brother's little boy, whose reindeer are those tended by Marrk.

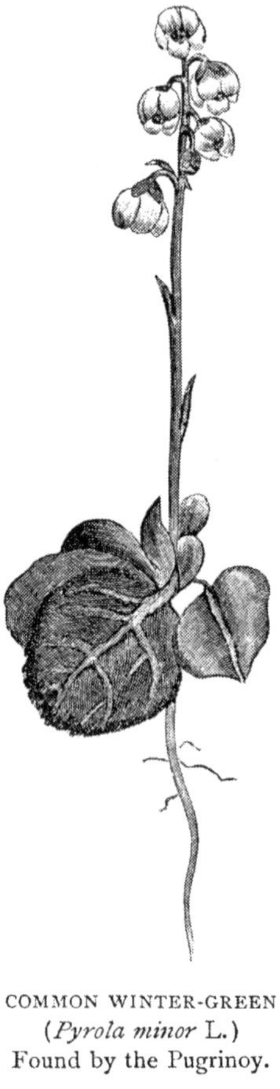

COMMON WINTER-GREEN
(*Pyrola minor* L.)
Found by the Pugrinoy.

But the feature of the Pugrinoy mouth is a very nice little chapel which was set up many years ago by Alexis (who owned the 10,000 reindeer).

Now Mekolka had come with us, and he, On Tipa, and Alexander spent a whole hour in lighting candles before the 'ikona' and in getting the incense to burn. The candles, which were stuck on by their melted bottoms, kept tumbling down, and then Alexander would cross himself many times and say a prayer of apology, the two Samoyeds standing behind him like acolytes, and copying his every movement.[1] At last all went well and the place was thoroughly censed.

[1] There was something singularly grotesque in this Rimmon-worship—something quaintly child-like in the idea of these poor Samoyeds, with their little idols all the

So we left.

Now Mekolka was driving a team of whites. And presently his Niasminta turned rusty and lay down. I thought the poor beast was tired, and said so to Alexander, but he answered, 'No, he is not tired, only he is bad.' That being so, he was a very bad reindeer indeed. For, do what we could, he wouldn't budge.

Then Mekolka turned him over on his back. You know that if you turn a beetle or a tortoise over on its back when it is 'shamming dead' it will come to life and try to turn over. Well, the white niasminta did the same, and jumped up on to its legs. But as soon as ever you tried to make it move down it dropped again. Many times we did this, but it always collapsed. So then they passed a thong round his neck, and the team dragged him along for, I should say, a hundred yards. I thought he must choke, but he didn't, nor would he come to life. After this we turned him over again, and when he tried to rise we prevented him, and so on many times. At last up he jumped and stood. We were afraid to try and drive him, so we tied him to the back of a sleigh and towed him home.

August 29th.—'Great bartering going on to-day. All the inland Samoyeds came down and exchanged their

while hidden in their clothes, going gravely through this function. I really believe that in their poor confused way they felt they were doing their best to keep in with both the Powers.

skins for tea, snuff, stuffs, and the girdles, or rather garters, called "poi-ess," made of woven crimson wool.'

August 30th.—Yesterday's gale raged all night, and the rain fell in torrents. It really beat us. The whole tent was flooded. It is useless to dig a trench round these Whymper tents, because the rain comes clean through the canvas of the walls. So we moved up into the empty isba, which belonged to Alexis, Alexander's cousin. Here I spent three or four hours in plugging up the cracks, and greased a bit of paper, with which I covered up the open window hole. I did not seem to feel the wet and cold, but Hyland complained always, feeling pains in all his joints, and was either in bed or shivering over the fire outside the huts, and I feared for him rheumatic fever. However, we got reindeer skins and made ourselves pretty comfortable in a dry corner.

They were shooting young bucks all day, and were very anxious that I should help, but after shooting three I had had enough. It was poor work. They shot seventy-five to-day, and the bellowing of the cows was most distressing. The women slipped off the skins and cut up the beasts ready for packing with remarkable rapidity.

August 31st.—Again we had rain and a gale. I was out very early, long before any one else was moving. Marrk drove all the herd back to the Pugrinoy last night to feed. But here were nineteen poor cows come back by themselves, wandering about and lowing piteously. To-day the remainder of the herd was brought back, and the twenty-five calves shot or caught with the di-zha.

Alexander's second man, Andrè, brought an old broken accordion, a 'garmonia,' he called it, into the hut, and played frightful tunes on it. Then Yakoff and he sang discordant nasal songs to it till the noise was truly diabolical.

September 1st.—We had a fine fire outside the huts, and I spent the greater part of the day drying our wet things, a work of some anxiety, as one had to hang them right in the smoke, and if not watched the sparks took hold of them. Hyland looked so wretched, creeping about the fire, that I chaffed him at last into taking a walk. So he went up by the creek, and returned in an hour saying he was warm, a very unusual case with him.

The three brothers Bulchikoff called on me to-day. Walking straight into the hut they sat down in silence, taking snuff. I went on with my writing, and not a word was exchanged for probably half an hour. At the end of that time I closed my book and asked to what I was indebted for the honour of this visit.

'Bolvana gorah' (Bolvan mountain), said the eldest son.

'Yes,' I answered, 'Bolvana gorah. What about it?'

Then it turned out they had remembered how old Marrk had taken me to Sieycherhur, and that they thought I might like to go to see Bolvana mountain. To this I agreed. We would go to-morrow. I would sleep in their choom, and the following day we should go on to see the bolvans.

I was very much astonished at this, but it seemed clearer when Alexander looked in for a moment and explained that these Bulchikoff were exceedingly poor, that they had looked to do a little trading with his cousin Alexis, who now, it seemed, must have been lost in the ice; that they hoped, by taking me to their choom, to catch me for a bit of trading. So I made it clear that I should not trade, but that, if that was it, we had better make a bargain at once. I would go, I said, on one condition, and I would pay them five roubles (for it was a very long journey, and it did not seem fair to do less). My condition was, that they should give me, or sell me, a little bolvan, one of those they wore under their clothes. Now, at this juncture old Bulchikoff, the father, came in.

'Bolvan,' said he, 'my son no bolvan. My son Christian.'

They were always suspicious of me, because of the Governor's letter; they connected me somehow with the ruling powers, and feared I would report them for idolatry.

'Nonsense,' I said; 'of course they have bolvans. Every Samoyed has bolvans.'

'My son's bolvan buried! Yes, yes, buried. No good,' persists the old man.

Oh, very well, I told them, then all engagements were off.

I need not continue. Those who know anything of the ways of savages or native people can fill in the details for themselves. It was a long palaver. Then I said good-day and turned to my writing. I suppose I wrote on for half-an-hour in a dead silence, then they began to mutter among themselves, and the old man left. At last it seemed they had come to some agreement. The youngest son—I saw it out of the corner of my eye—fumbled about in his clothes and produced a little coloured doll, which he brought to me, and said, 'I good Christian. My bolvan, your bolvan. Yes, yes.' I looked at it. There was a shiver of protest when I raised the hood from the face. It was apparently quite new, but he said he had made it twenty years ago. 'No,' I said, 'too clean; I want dirty bolvan.'

Well, after an incredible time and much jabbering, the eldest son brought me a second bolvan. There was no doubt about its dirt or its antiquity. The face, although it had always been covered up, was so worn that the features were all but lost. The wood, originally light coloured, was now quite black. He told me that his father gave him this bolvan thirty years before; that his father's father had owned it, and he did not know who first had made it; but it was so very old, 'oh, more than a hundred years!' 'A hundred' with these people means some immense indefinite age, past all calculation. He said it had had many dresses, so many had worn out. I could believe it all; it looked like some little dirty old mummy. Well, they struck a bargain, they were so hard up, and I got my bolvan, and my friends left. The picture gives a fairer idea of this than any words.

THE OLD PERSONAL BOLVAN: IN AND OUT ITS CLOTHES

At tea this evening Alexander offered me as a *bonne bouche* the boiled eye of a reindeer. He was astonished when I refused it, but not disappointed, for he ate it instead.

Sunday, September 2nd.—At last a six days' gale has moderated. As it is Sunday, the Russians lit up candles in the chapel and burnt much incense; the poor muddled old Samoyeds going through most wonderful contortions and genuflections behind Alexander's back, for he officiated. Then we cooked pancakes of flour and water, and thought them very good.

I find that I never told you of our old cook Anna, Anka, Anaka, Ananinka—for she answered to all these names according as you were angry with her or pleased; Anna being the term of command, Ananinka of wheedling or entreaty, and the other means between the two. She was Mrs. Uano's eldest daughter. She was a widow, and lived in the service of Alexander at Okshin on the Petchora. An old woman in appearance (though she could not have been more than thirty),[1] she ruled the fire and the pot, and I found her kindness itself. There was nothing the old creature would not do for Hyland and myself, and she was always full of laughter at our ways. Every year she crossed to Kolguev to do Alexander's work and see her friends.

In the evening I left with the Bulchikoff brothers. Their choom was only ten versts away. The first five versts we did in twenty-five minutes, the last in thirty-five minutes; for the first we did at the lope or canter, the next at a slow trot. This will give an idea of reindeer pace in summer.

Now I do not claim to have been very particular, but I must confess I was disgusted this time. Such a choom as it was! Dirty, full of holes and very small, it could not have been more than twelve feet across, for when I stretched myself out to go to sleep my feet were in the fire; so we had to send the boys out for water to

[1] People such as these, with whom the women are grown-up, in a sense, at about thirteen, age so rapidly that a woman of twenty-five looks old.

put the fire out. In this wretched place were many dogs and eight persons besides myself, namely, the four brothers (a fresh one was there), the eldest son's four sons,[1] and a little girl. For supper we had ripe cloud-berries in tea, and boiled reindeer. They offered me as a great delicacy some cold liver, but, as I never touch liver, I refused. The man's wife was dead, and, as always happens when there is no woman, the family were simply 'pigging it.'

After the 'salnik' or lamp was lit (this was a bowl of reindeer fat in which a bit of old fish-net served for wick) they sang sagas, and then we lay down for sleep. But I slept very little. It happened that my feet were over some old fish-bones, at which the dogs kept pulling. After a bit I let out, and caught one of them a hearty kick. Whereupon the dog, believing itself attacked by its fellow, set instantly upon it, and a furious fight ensued. Also, no sooner was the light out than it seemed a whole Noah's ark of fierce invertebrates were loosed upon me, and the attack went on till morning.

September 3rd.—As a consequence of all this I was moving earlier than any one else. The bitter gale of yesterday had died out, and the ground and all the water was bound by a hard frost. We had breakfast

[1] Like all Samoyeds, the boys had a Samoyed and a Russian name apiece—

1. Ludi (Samoyed), Roman (Russian).
2. Nensika or Petrooska.
3. Mozük or Mekolka.
4. Tierrki or Simeon.

at 7.30, and then left. For breakfast we warmed our last night's lamp, pulled the wick out, and then ate the grease with black bread. It was not a *recherché* meal, but it was economical. Then Abeiloh, the eldest brother, left with me for the bolvan hill.

After a journey (uneventful except that I picked a harebell bloom, and that we passed the Bulchikoff's winter seat — an ombara or hut, set in mouldering relics of Samoyed dirt of every description) we reached Bolvana Mountain. Sahbor — the cathedral — the Samoyeds call it, using the Russian word.

Here I found nineteen bolvans or idols. They were not stuck up, but were lying side by side in a row. A few were comparatively new, but the majority were very, very old, and grey with lichen. All had the same features and the same ribs. Also there was a little broken sleigh, a spoon, and the remains of a flour-tub —propitiatory gifts from dead men's friends. I was naturally very anxious to secure some of these, but found it difficult.

Abeiloh's behaviour was very different from that of old Marrk. Instead of lying prone he came with me to the very spot, and there, standing face to the sun, with much bowing and extending of the arms, he ran on in a continuous stream of words, of which I could not gather any sense. But I felt he had his eye on me all the while, for whenever I moved a hand towards the bolvans he turned quickly round with a cunning intelligence on his face. But I was one too many for him. For I offered him a cigarette which I had made from a leaf of my note-book. And while he was stooping down to light this I managed to slip a small bolvan and the spoon into my pocket unperceived.

On the return journey my sleigh caught in a hummock, and the whole of one side broke off. But Abeiloh had

BOLVANS OF THE HOLY HILL

mended it in fifteen minutes with seal-thong lashings, and we went on.

We reached the ombara, or hut, again in a blinding snow-storm. But here I picked up a capital little adliurs, on which I could sit astride, and came along in fine style, riding over all the hummocks delightfully.

I had had enough of the Bulchikoff brothers and their choom, so went straight away back to Scharok. Here

I found Hyland better; and he accepted an invitation from Verrmyah and Katrina to spend the night in their choom, now pitched by the Baroshika. At Scharok also I found the whole of the Uano family, who had taken up their abode in a small lean-to. That day seventy-six of Uano's calves had been killed, and Hyland told me that four men had skinned seventy-eight calves in six hours; which, said Hyland (who had had some experience in that line), was very good work.

There was great rejoicing in Alexander's hut; for at the harbour mouth had appeared a sail, which could be no other than Alexis'. So he had not died after all. And this was proved later on when the vessel, having come to an anchorage in the creek a mile and a half away, some of his men came ashore. It appeared that, since his departure from Okshin in June, he had been slowly coasting along, bartering with mainland Samoyeds.

This evening I came across for the first time a character destined to play (though we did not know it then) a considerable part in our story.

I was sitting in Alexander's hut drinking tea (Uano was there, and Mrs. Uano, dirty Shabla, who had come up from the Gobista, and one or two more) when suddenly the door opened and a most extraordinary figure appeared.

No words of mine can possibly convey to you even

a remote idea of this man's astonishing appearance. One eye was large and white and sightless, the other rolled round and round, as though it were attached to a stalk, like the orbit of a cuttle-fish. His retreating forehead, with prominent gorilla-like orbital bones, lost itself beneath a skin hood, from underneath which appeared long locks of straight black hair. His mouth, enormous and wide open, was hedged by a bristling row of ape-like teeth, and altogether there was something so ogreish and strange about this apparition that I stared speechless, at which they roared with laughter, and then arose a shout, 'Tima Fè! Tima Fè!' It was Tima Fè, a Samoyed from the other boat. He grinned at his reception, and when he grinned he really seemed to lose himself behind his mouth.

This Tima Fè was sort of henchman to Alexis, and had been sent ashore to get the isba ready. He sat down, was given some vodki, and then we became, as usual, the subject of an up-and-down conversation. Tima Fè listened to all they had to say about us; how we had lost our ship, how we had lived with the Samoyeds, and how we were hoping to get to England, which was an island lying beyond Norway right across another sea. On all this he made no comment. But when they had told him all they could he was silent and pondering for a bit,

and then, turning round, looked me straight in the face for quite a while, and then in a hoarse cracked voice, like some old crow, said twice over and very slowly, 'England, impossible! England far away; far, far away.' (*Ahnglia nailzah! ne mozhnah. Ahnglia dalyoko, otchen, otchen dalyoko*), and then relapsed into silence.

There was some touch in this man's face that told one at once that he was not all there. He was, in fact, a bit soft. An excellent workman, a recognised authority on many matters (they all accepted him as the weather-prophet), he was one of those curious twisted phenomena one sometimes comes across, where cunning is combined with warped wits. He was a natural, and a very uncanny one at that, with a curious influence over all the Samoyeds. I think even the Russians in their hearts feared this strange creature, for they were always most careful not to round counter to him in any way.

September 4th.—'NE. Bitter cold. Snow. Gale again. Young Uano—"Uano droogoy," as he calls himself, who sat long in tears in my hut yesterday saying his Russian was dead, drowned, and he himself was broken—rushed in this morning in great excitement, seized my hand and cried, "My Russian here, my Russian! my Russian!' as if he could not say it often enough. So that, what between his tears and laughter, for he was quite hysterical, I had all my work cut out to humour him and get him gone. But his Russian, Alexis, did not land to-day. Instead he sent Yugor, a curly-headed, heavy Samoyed, to help Tima Fè in his work.

'Hyland returned from Verrmyah's choom, having had great fun in driving "adliurgo" with dirty Shabla after the herd, which they brought down. Our hut was miserable lodging to-day, for the snow soaked in.'

September 5th.—'N.NE. Gale all day, with snow and rain. Bitter cold. The curly-headed Yugor came drunk into my hut, and after turning him out twice (he was all liquid mud from head to foot from falling in the creek) I had to fire my collecting gun over his head, which scared him so that he crawled away crossing himself in terror. Yakoff Popoff brings in word that the karbass (boat) is broken by swinging round on to a stone. Alexander, much concerned, says it is the fault of my bolvan.'

(This was the first sign of a grievance which developed unpleasantly later on.)

September 6th.—E. Rather warmer, with gale and driving rain all day. The whole area on which the huts stood now resembled a slaughter-house. As you walked about you had momentarily to creep under lines of flapping wet raw reindeer skins; the ground was drenched in blood and covered with reindeer paunches and skulls. But I think the diary itself will give a sufficient idea of this.

OUR QUARTERS AT SCHAROK

Face page 347

'To-day we cleared out of Alexis' isba (dwelling-hut) into a little ombara (storage-hut). A certain amount of wet is kept out by a polar bear-skin stretched over the top, but still the wet comes in so that the planks or beds are all wet. However, we did our best by packing the cracks to keep it out. There is no window, but a hole over the door, and the approach is filthy; guts, blood, etc., lying all round the door, so that Yakoff and I slipped and fell as we entered. But I have brought a lot of moss, and have paved the front with this, so that it is better. Yet the smell is most unpleasant, and I am thankful the weather is not hot. Supporting the planks of our beds are barrels filled with I don't know what, but it smells horribly. The tent is still standing all wet. I spent a good deal of time down there to-day, trying to get things straight and emptying water out, but it rained hard all the time. Alexis Samarokoff was towed in to-day. He is a fine-looking Russian. He tells me the Governor of Archangel, travelling to the Petchora in June, was stopped by ice at Mezen and compelled to abandon his ship and travel overland to Pustosersk, and that Sibiriakoff brought him thence in his steamer.[1]

'Gave Alexis the last drop of whisky, which has remained all this time in my flask. He said it was wonderful, and rubbed his stomach long after it had gone down.'

[1] Which I found afterwards was a fact.

September 7th.—'NE. Blew hard till mid-day, when abated, and evening was calm. With the exception of a slight abatement on Monday last we have had ten days of gale—not wind, but *gale.* This has been a woful day of Samoyed drunkenness. It is a point of honour with them to drink fair. And as old Yelisei was unable to move from his position face downwards on the ground, the women turned him over and poured vodki down his throat. There were, of course, comical interludes, *e.g.*, an old woman crawling inch by inch up the steep plank to Alexis' karbass. Sketched chapel and our ombara, but the wind was biting. The stielka grass is only now coming well into flower.'

DRESS OF THE RUSSIAN TRADERS

September 8th.—'W. Quiet in A.M., rain and wind P.M. Alexander's karbass has swung on to the mud; there was no water at high tide, so another good day has been lost. It is very trying to see all this favourable wind going by.

'Alexis' men killed a hundred young deer to-day,

taking with the di-zha those the Samoyeds did not shoot. The Russians are far superior to the Samoyeds at this game. The Samoyeds use an underhand toss, but the Russians throw overhand in cow-boy fashion. It is a queer picture. You have the big herd divided into two, a constant stream of cows and calves bolting from one side to the other. The Samoyeds are squatting here and there on one knee with their old flintlocks, and the Russians creeping low about with the di-zha ready coiled. The dogs bark, the cows and calves call or grunt incessantly to one another, the cows always trying to screen the calves by keeping them close at their side. Every now and then a calf goes down, and the mother stops for a moment smelling it and grunting, but the calf makes no response, and she is soon hurried on by the press. She will come back there presently and try to move it with her head, and finding it does not stir she stands over it for a moment or two, a sad picture of wild sorrow, and then walks slowly away, stopping at every few steps to call it on. At first the deer keep breaking and trying to bolt, but, being always headed by the dogs, give it up as hopeless. All the same, every now and then a few will get together and creep off. If they manage to get a hundred yards without being noticed they then stop, lay their horns right back on their shoulders, break into a swinging trot, and then into a gallop. When they mean going like this they sometimes succeed in getting clear off; and they would always, if they could only keep a straight course and act decidedly. But too often, poor beasts, they shift from point to point, and then the adliurs (which to-day was entirely in the hands of Marrk's old wife) goes off with the dogs, and heading them by creeping up some hollow, brings them back.

'To-day a splendid young white-faced buck quite defeated them. The five men all laid themselves out to take this beast. But he seemed to bear a charmed life. Time after time they threw the di-zha over him, but always he managed to escape. He trotted along right in front of his mother, and did move splendidly. At last they said he was not a reindeer, but the devil, and gave it up. I shot two purple sandpipers. A couple of turnstones spend much time at our ombara, turning over the deer's paunches.'

Sunday, September 9th.—'W. to N. To-day we have had sun, snow, rain, hail, and frost. Everywhere on the tundra the ice is now thick, and the edges of the creek are frozen at low tide. The boat was floated off to-day, but as it was Sunday, they said it was not right to finish storing the barrels, only they might kill reindeer and drink vodki; which they did. I went off in a boat to the outer koski with Alexander to take observations of the harbour. He was much interested in the prismatic compass, which he could not understand. While the men were taking the young bucks to-day one of them, caught by the horns, snatched the di-zha from

André's hands, and bolted into the sea. They got two boats out after him, but to my great delight he swam splendidly, clean distancing the boats against the tide. When I last saw him he was galloping over the shallows for Barakoska Nos.'

September 10*th*.—'S., strong wind and snow, with hard frost. Alexander's karbass ready.'

September 11*th*.—'S. to S.W. Frozen up. Snowing.

'I spent much time away on the tundra with the axe cutting sections of moss heaps which much interest me. The Russians would not go out to-day, as they said it was not right to move on the Tsar's birthday. This evening, after a day of idleness and vodki, they have been burning seal-fat barrels and shouting hurrah! They put a 5-lb. tin of gunpowder into a barrel which they stood on a bonfire. It went off and blew bits of burning wood all over the huts, but no one minded.'

September 12*th*.—'S., SE., E. This has been a day of really tremendous rain. I have spent the day in fighting the water with rags and tins, but have not found a single dry spot to write in. Hyland went to bed in the afternoon feeling sick—probably from want of vegetable food. I also physicked Alexander, who complains of pains and chill.'

September 13th.—'NE. Strong wind with rain showers. (It may be worth noticing that the barometer in the last three days stood 29·8, 29·3, 28·9, 29·5 in. These extraordinary fluctuations boded no good.) Alexis, who also has an aneroid, says he cannot understand what I mean by saying it won't be fine, for the finger of his glass points, he says, to "fine." I tried, but without any success, to explain to him how we calculate from relative rise and fall. His answer always was that if the glass said "wet" it must be wet, if it said "fine" it must be fine, unless the glass was broken.'

September 13th.—'Now both the karbasses are ready. Their holds are filled with barrels full of reindeer meat and the seal-fat from last winter's catch. Over these are spread reindeer and seal-skins, and the pelts of two walrus (all they have taken), and skins also are hanging from the rigging to dry.'

Alexis' karbass was rather smaller than ours, but the rig was the same. My sketch will explain this better than any words. So at six o'clock on this evening we made a move, and sailed out till we were just inside the entrance to the harbour.

We had a visit from Verrmyah and Katrina just before we left, and divided between them and their parents such little property as we could spare—to Uano our saucepan (the only cooking utensil we possessed), to Katrina the red flannel flag (for Adski), to Mekolka the axe, and so on. Also I gave Kallina his knife. He had quite made up his mind that as our boat had not come he should not get the knife, and was saying something sorrowfully about it when I whipped it out and gave it to him. You should have seen his face change! I was frankly sorry to leave these kind friends, to whom we owed so much, and the feeling, I know, was shared by them.

ALEXANDER'S KARBASS OR BOAT

'Good-bye, Kolguev!' shouted Alexander as we moved off. He had long been practising this bit of English under Hyland's tuition, and Hyland shouted it, too, gaily enough. He was only too pleased to be off. We all cheered. But I knew very well that we were not gone yet. Everything in my view pointed to bad weather. And at least one person took care that we should not plume ourselves too soon. Tima Fè, perched on the cabin roof or poop of Alexis' boat, lay to with his tiller for a moment as he croaked back his answer to our cheer.

'England far away. Far, *far* away,' like the long-drawn howl of some old prairie wolf.

The boat was thirty feet long, with a ten-foot beam, nearly flat-bottomed, and drew six feet of water when laden. In the stern was a little cabin, in the centre of which was a table; on either side of which we lay—Alexander and Yakoff, Hyland and myself. It was only constructed for two persons, but with movable planks we managed for all.

In the middle of the boat was a hold, where, above the barrels and skins, slept the five Samoyeds who formed the crew. Here, also, when she was not cooking, lived old Anka, always sewing away at skins, and with her a little red fox. She had also kept three charming little blue foxes at Scharok, but all had died.

There was little to do on the boat, so I made a pack of cards, with which Hyland and I played Nap.

September 14th.—NE. A very wild day, with incessant heavy snow. These boats can only run before the wind, so with this wind it was useless to think of moving. For the idea was either to make for the channel behind Sengenski Island, or to run for the Petchora mouth.

'There was much excitement this morning, for the karbass dragged her anchors; but after a bit we got out a third and steadied her, narrowly avoiding a collision with the *Michael* (Alexis' boat).

'*September* 15*th.*—To-day we had brilliant sunshine, and at noon I was able to get sights and correct our watches.

'We had to send on shore for water to-day, and Hyland went with the boat. The snow was very deep, and many tracks of foxes round the huts. Immense

ALEXANDER'S KARBASS
In which we crossed

number of glaucous and herring gulls, and one fulmar petrel—the first I have seen here near the land.

'To-day Alexander said, "This bad weather is because of your bolvan. Throw it away." To which I answered "Gammon" (poostoi).'

Sunday, September 16th.—'The morning broke clear with a strong wind, which increased in afternoon to a gale with snow. To-night the gale is *raging*. In the morning Alexander, Yakoff, Hyland, and I crossed in a boat to the sand-banks, and walked down to our old sleeping-place. A blue fox was hunting for molluscs along by the point. Hyland, who had his gun, tried to intercept it, but without success, for it swam straight across the mouth of the river and escaped. We saw many flocks of king eider and waders, chiefly sanderlings and dunlins. I collected sea-weeds and hydrozoa from the gulls' nests, and picked up several nodules of stiff rounded ooze, which formed with stones and shells an elementary conglomerate, such as I have found in a complete state elsewhere on the island. The storm increasing, Alexis' karbass dragged her anchors—we had to send four men over to help him—and bore down on us so rapidly that we only just escaped a collision. Alexander has been a great nuisance all day. He talks and grumbles persistently about my wretched little bolvan. "If you will burn it I will get you many from the Timanskii tundra," he says. But a "Timanskii bolvan is not a Kolguev bolvan, and a bird in the hand I stick to," say I. But there is growing dissatisfaction about this; for the Russians and the Samoyeds sit and talk about it eternally. Both are afraid of it, the Russians because, being Christians, they have bolvan in their ship, and the others because a "Turk" has one of their gods.[1]

[1] For when the Russians were catechising me about our religion in England they

'My glass has gone up so steadily from 29·2 in. to 29·6 in. that, taking that and the sky together, I felt I might venture on a prophecy. "To-morrow," I said, "it will be fine. Let the bolvan be." And for that time they were content. Only it blows at present as hard as ever, and whenever I go on deck down the wind comes the mocking long-drawn howl of Tima Fè, "England far away. Far, *far* away!" till I could shoot him for a raven.'

September 17*th.*—'NW. Calm. Fog. Bar. 29·7 in.' So I had scored up to a certain point; the wind was fair but the fog was thick. Alexander, who burnt much incense to-day, seemed to have forgotten the bolvan. So they hold this poor little idol responsible for storms, but not for fogs. Reputations are easily won. I seem to be suddenly established as a good weather-prophet. They actually came and asked me to-day whether we should be able to leave to-morrow. It is pretty risky to prophesy about Kolguev weather, but thinking that, if once we started, we should get to land somehow, and that we knew as much about it as we were likely to know, I answered boldly, 'Certainly.'

September 18*th.*—(Michaelmas Day, old style). Foggy, with the barometer at 29·7 in., but showing a tendency

had said, 'Do you worship St. Nicholas?' 'No.' 'Ah, not Christians.' 'Oh yes, Christians.' 'Why, how is that? Have you God?' 'Yes.' 'And Christ?' 'Oh yes, of course we have Christ.' Ah,' said Yakoff, 'Turks.' This was a grand discovery; after that we were always Turks.

to fall. The wind to-day boxed the compass, setting at mid-day N., and finally again at NW.—our best wind. I wish I could give you a phonograph reproduction of all that was said to-day. From early morning till mid-day there was a continuous conversation shouted between the two boats. I knew there was not the smallest use in talking; I never argued with these people. I said my say and settled down to my writing. This was the sort of thing:—

ALEXANDER (*from the poop of our boat*). Much is wrong. Nicholas' lamp keeps going out.

ALEXIS (*from the poop of his*). The weather will be terrible. Much fog. I do not go to-day.

ALEXANDER. But I am weary. I want to see my wife. But this Kolguev is a bad island. I come no more. My son comes.

ALEXIS. There will be a great storm. The wind is NE., and my karbass likes it NW. I do not go. What does the Englishman say?

ALEXANDER. He says he thinks at sea no fog. The wind and the sun go together. When the wind northwest then we must go quickly.

ALEXIS. Does he say that?

ALEXANDER. How? I cannot hear.

ALEXIS. Does—he—say—*that*?!!

ALEXANDER. Say what?

So then they would have it all over again till they were out of breath, when Alexis would descend to his deck to confer with his Samoyeds, and Alexander to his deck, to confer with his.

Presently they would be up on their poops and at it again.

ALEXIS. Where is the Englishman?

ALEXANDER. In the cabin.

ALEXIS. I want to see him.

ALEXANDER. He writes.

ALEXIS. Hah! Always writing. (They looked on my writing so much as part of my uncanniness.) What else did he say?

ALEXANDER. If you don't go to-day you may not get off at all.

ALEXIS. But this bolvan. Bulchikoff had a bolvan on board last year and the ship was lost. You know it.

ALEXANDER. I know it. But he does not burn it, though I speak to him so much. Many times.

ALEXIS. But my glass down goes. What does he say to that?

ALEXANDER. He says his glass down goes, but no matter, so slowly. He says it must go down or up if it is not still.

ALEXIS. Does he say that? Well, that is spoken very well.

And so on a hundred times over.

You may imagine how I laughed at all this rubbish as I lay in the cabin. But I did not feel happy, though I was quite determined to get them off if I could. I knew that if they did start on my advice and the old karbasses were to get into difficulties, they would round on me as a Jonah. But I was sick of this. So presently, when Alexander came down into the cabin and said his cousin wouldn't go, and he couldn't go without his cousin—what did I think? I just looked him in the face and said as follows :—'Alexander Samarokoff, I consider you a fool (glupoi). You say that you have always been able to find a good bit of weather by waiting, and that you have never known the winter set in early like this. What has that got to do with it since it *has* set in? If you don't go now—and you have a wind which will take you quickly that little run—you will deserve to get caught in a trap. How do you know that after this we may not have a ten days' easterly gale on shore as we had before? your wife may whistle for you then.'

He sat as though struck. All of a sudden this reasoning seemed to catch hold of him; he rushed on deck, shouted out to his cousin that he was going with or without him. Then Alexis surrendered, for he would not be left behind.

So Alexander called all hands into the little cabin, where they stood jammed tight (he turned poor old Anna out 'because she was a woman,' he said) while he said prayer after prayer to St. Nicholas' ikon (image), and swung the censer till the poor Samoyeds coughed and choked again. Then out they all rushed shouting, 'Now away, away!' as if bound on some valiant quest.

On the wood-work of the shelter of the companion were deeply carved in block capitals the English words TESS OF CROMER. Alexander could tell me nothing more than that it was part of a bit of wreckage picked up by himself on the Kolguev sandbanks twelve years ago. Since my return home I have done my best to trace this boat, but without success.

But to return. Alexander after this function betrayed but one more sign of irresolution. How could he enter the Timanskii Schar at night? 'Well, you must hang about till the morning,' I replied, 'but you ought to have lanterns at these entrances, as we in England do.' Whereupon says Yakoff, 'Ah, he's getting frightened now; he says there ought to be a lantern.' And they all took it jeeringly up like a stage chorus, showing the curious nervous state into which they had worked themselves.

So we weighed anchor and moved out one behind the other. Over the bar we found eight feet of water, and got safely out. Very soon we were clear of the fog, which we could see, whenever we looked back, hanging like a pall of cloud over the island. We had done it. Kolguev was behind us.

Kolguev, with its kind hearts and its barbarous contradictions, its many delights and many discomforts, its treeless wastes and its charm of birds, its reindeer, its sleighs, its careless existence—all these were things of the past.

This time it really was 'Good-bye, Kolguev!'

CHAPTER II

THE LAMP OF ST. NICHOLAS

In the cabin hung three lamps before three ikona. On the left St. Philip of Solovetsk, on the right St. Alexander, in the middle St. Nicholas. These had each its lamp; but that of St. Nicholas was the biggest lamp, because his ikon was very old—(it had belonged to Alexander's father)—and he is patron to all who follow the sea.

This lamp was always the subject of much concern. But to-day something surely ailed it, for it would not burn. In vain Alexander stirred the wick with a pin, in vain Yakoff warmed the holy oil and fed it afresh. It would burn perhaps for half an hour, and then die down. The oil had thickened with the cold; that was obvious, but not obvious enough for Alexander. 'Your bolvan,' he said, 'is in the box right under the Holy Nicholas. He likes it not. Can his lamp then burn?'

Had this happened earlier I would have shifted it just to humour him. But now we were, so to say, pitted one against the other: I was sick of all the nonsense, and I vowed I wouldn't budge an inch. I knew that to move the bolvan would be the first step to losing it altogether.

But they tried all ways. They tackled Hyland when he was alone on the deck, as he told me. They told him he could get it if he liked. Of course Hyland was as staunch as a rock—he only laughed at them and said he knew nothing of it.

I nearly lost it once. I was feeling for something in my long tin case, the lid open, when something made me turn suddenly, and there, close over my shoulder, was the face of Yakoff, the flicker of the lamp playing on his features, and his eyes fixed on the contents of the box. He couldn't have seen the bolvan (for it was in a little skin bag), or he would have snatched it. At the same moment some one went out through the door in the darkness behind him. I knew it was Alexander.

Now the following had happened earlier in the evening before it grew dark. Hyland, who constantly had his gun on the deck for chance shots, had brought it down to the cabin, saying that Alexander had forbidden him to have it there. This he considered very hard lines, as all the Samoyeds had their loaded rifles. It was very unlike Alexander, who took much interest in this deck-shooting at seals. Well, my first impulse was to go and expostulate, but after all, I reflected, he is master of his own ship, so, though I thought it unfriendly, I let it go. But I had been turning it over in my mind, and though I could not see it clearly I understood intuitively that it was somehow connected with the contra-bolvan spirit. Now it flashed on me in a moment. They knew well enough that we never kept our guns loaded in the cabin, and doubtless they meant to have us at a disadvantage should occasion arise. I thought it best to settle the point. Yakoff couldn't go; he was cornered, as it were, so after his first surprise at discovery he tried to appear unconcerned, pretending he had come in about the lamp.

'Now, look here, Yakoff Popoff,' I began, 'you remember the Governor's letters with the seal, and what he said?' I smiled to myself as I said this, my voice sounded so innocent, like the quiet beginnings of Peter Simple's Mr. Chucks.

'Yes, yes,' says Yakoff.

'Good. Well, the Governor can send you and Alexander and the lot of you to Siberia, unless I save him the trouble.' With that I opened the box again, took out my revolver, slipped in six cartridges, Yakoff watching all the time, and put it in my pocket.

'I am a soldier,' says Yakoff with a fine show of courage.

'You may be a soldier,' I answered, 'for all I know or care, but you look very like a fox.'

I used instead of the Russian word 'lisitsa' the Samoyed word 'tchuornia.' You could not insult these Russians more than by using Samoyed words of them. Yakoff retired to the deck.

Then I thought I would make assurance sure. I took out my bolvans, the one they knew, and the mountain ones of which they knew nothing, and putting them in my pocket turned the key of the box. 'Now, my friends,' I thought, 'you may chuck the box over, but you don't get the bolvans.'

Of course it was all discussed on deck, but when I went up I found them all very quiet, and nothing was said.

Alexander and Yakoff both came down into the cabin presently, and talked with us as if nothing had happened. All seemed going well. The wind was a bit stiffer, but all in our favour, the night was clear, and St. Nicholas' lamp was burning well, for the cabin air was warm.

Hyland, who knew nothing of all this—I had not told him, and he did not understand Russian—soon went to bed, and we three went on deck.

The wind was certainly freshening, and we went along at a fair speed. After stopping on deck for an hour or more, I thought I too would turn in. I could just make out our consort's outline, and, holding the cabin door in my hand, I shouted out 'Good-night' to Tima Fè, who, as I knew, was at the helm. Was there ever such a Solomon Eagle as he! My voice roused him in a moment. His muddled old brain pitched on the only theme it could connect with me.

'Bad bolvan, bad bolvan,' he shouted back. And then across the water came the inevitable dreary formula, 'England far away. Yes, yes, far, *far* away.'

I stumbled down into the cabin. All was dark. The Lamp of St. Nicholas had gone out.

CHAPTER III

THE MAINLAND

September 19*th.*—When I came on deck this morning I found the wind high and a driving snowstorm raging. We were just opposite to Timanskii Schar. We were far in advance of the *Michael*, but, dropping sail, we waited for her. When within hailing distance they wisely determined not to attempt the Schar, but ran on some forty versts to the Kolokolkova (bell-shaped) gulf. We ran in here with fourteen feet of water over the bar, and after going aground and getting off again with the help of long poles, we dropped anchor. Nineteen hours and half since leaving Kolguev.

These long poles, divided into quarter 'arsheen' (seven-inch) measurements, are called 'pomorrna.' Many of them were on board. They are used for punting, and have at the top a button which is pressed against the chest.

A fine white whale rose right under the gunwale.

We celebrated our arrival by drinking vodki to the toast called 'Sprebuteum,' which always celebrates a voyage accomplished. Then they said I had great wisdom. I understood all weather, and but for me they would not be here. I wasn't so sure of that. It struck me they would gladly have come without me had they seen their way.

'Yes,' says Alexander, 'Kolguev is an island, and your England is an island. Very much sea. So you understand all ships and all islands, and that is why your Queen Victoria—your woman king—rules all the islands of the world.'

So we were friends again. Only once was there any reference to the bolvan after that. Just before they went to sleep Alexander and Yakoff came to me, and smiling said, 'Look in your box. Your bolvan is not there. The Holy Nicholas there; your bolvan not there. Look and see.' What on earth they were driving at I never could guess. Perhaps they had been exorcising it. They had expected an indignant denial, but were disappointed. For my little bolvans were all safe in my pocket, so that I could honestly answer, 'I needn't look, I know it. Quite right, the bolvans are not there.' It took the wind out of their sails, and much astonished, while it flattered them in some way, for they beamed with delight and went to sleep.

In the distance we could see a choom on the river bank, and a man walking along.

Hyland was in high spirits at our safe arrival. He looked forward with great delight to his return to the little business which he had started in England a few months before he came away. Other and more tender considerations also weighed in his case, but these I will leave to the reader's fancy. Old Sailor, too, though he had had the run of the boat and had slept at night in our cabin, was in high feather at the smell of land. His stump of a tail wagged incessantly, and, poking his old nose out of every hole in the gunwale, he indulged in sniffs as prolonged and sounding as though he were some old snuff-taker.

Though we were now within two miles of the land, we were not to reach it till six days later.

The gale blew hard all day, and we could do nothing. This was very trying, because if we could not get quickly to the Petchora (and we are now lying only a hundred and twenty versts from Okshin) we might there find a salmon-boat which would run us to Archangel in time to take the last English trader before it left the port on October 1st.

The gale died out towards 9 P.M., and then, as to-morrow is a feast-day of St. Mary, Alexander filled the cabin with incense till we all choked again.

The next day was quiet but disappointing. For two Samoyeds crossed in a little boat from the left bank and told us we should find water enough higher up, so that we could land on the right or southern side of the gulf, find chooms and reindeer, and go straight to Okshin.

We sent the boat out to take soundings, and they returned to report plenty of water. So at noon, with the rising tide, we began to punt the karbass up one of the river channels. For two rivers enter this gulf, viz., the Kambalnitsa and the Kolokolkova. But we grounded so often that we gave it up, and with great difficulty worked our way down again, and so lay with a foot of water under our keel.

One of the mainland Samoyeds, whose name was Pronia, crossed again, and returned bringing his wife to call upon our old Anka. They both rowed, with the baby on the bottom of the boat between them. I was much amused with the women's quaint politenesses to one another. Mrs. Pronia wore earrings with bunches of hare's fur in them. These people brought kambola, which our crew ate raw. They told me there are no salmon here, but admitted they had never set nets to see. On this evening, as our fuel is reduced to two logs, we went by invitation to supper with Alexis.

On our return our visitors dropped the baby into their tiny boat (which a man could easily lift on to his shoulder) and pulled off across the dark river-mouth with a two-mile journey before them.

The following morning opened fine, and Alexander had quite made up his mind to run out with the tide and make for the Timanskii Schar, behind Sengenskii Island, which is only a hundred versts from his home. He has never before been in our present harbour, and does not like it. But just as we were off a fog came on, and then the wind chopped to the north, and we were prisoners.

So there was nothing for it but to warp the boats to

the shore. We worked very hard, Alexis' men coming to help. We sent the boat out to drop a kedge, and then some hauled on the line while others punted with the 'pomorrna.' I suppose most of us are happiest when we have real hard muscular work to do; when one can thoroughly let oneself out, so to say, until one can't move another finger. These poor fellows all worked in silence, so I started a chant, which they tried to follow till we were all howling different tunes like demons. Then I wove the days of the week into a rough sing-song and cheered when Sunday came, and, poor chaps! they got very enthusiastic about this, and pounced down on Sunday with such a squall, you never heard anything like it. But this inch-by-inch work is very severe: we only made some half mile, if that, and at last the wind raged so wildly that eleven men on the rope could not move her half-an-inch, and it really was not safe to send out the boat. So we stayed.

Alexander was fairly astonished, and said more complimentary things about Queen Victoria, and her islands, and her wonderful Englishmen, than modesty will let me repeat here. Frankly, I was glad to feel that all past unpleasantness was now wiped out.

We did manage to get our boat off to the other karbass, but she could not return.

And the following day a northerly gale raged, with pelting rain. The capstan creaked so badly that we feared it would go, and so the chain cable was wound round the mast. And all day, with three anchors out, we could only just hold our position.

So Sunday came, and with it the gale from the north continued, but with a furious blizzard of small hard snow. The fire was usually made in an iron pan on the deck, and as it couldn't be lit, we, Hyland and I, came badly off for food. For, unlike the Russians and Samoyeds, we did not care for raw stale reindeer flesh.

Also, as it was Sunday, little candles were lit in front of each ikon, and before the pictures of saints, of which there were three. They were one candle short: and it was very funny to hear Alexander calling to his men, 'Hurry, hurry, Yakoff, get a candle! All the others have their candles, but big Nicholas has not got his. That is not good, not good. More storm.'

Monday broke with rather less of a gale, though the snow continued unabated and the sea ran nearly as high as ever.

Again we worked at warping. Hyland, whom I would not allow to work yesterday because of the snow, took it in turns with me to-day, for we had only one tarpaulin coat between us. He worked spendidly, and his weight and strength were of the greatest service to them. In that respect he was of far more use than I, who cannot pretend to much strength of that kind. The plucky little Samoyeds were beyond all praise, for the decks were slippery with snow and wet, and their seal-skin malitsas and their soviks were soaking through.

It was well on into the afternoon when, having at last cleared one long straight channel in the teeth of the wind, we were able to set our foresail and mizzen, and in fifteen minutes we had run in under the lee of the land. Then the hands went ashore, and Hyland with them to tow, while Alexander took the 'pomorrna,' I the helm, and so we moved up a mile until we were under Pronia's choom, in deep water close to land. We took the poor little fox ashore, who straightway made himself a capital kennel in the bank. Old Sailor, too, was mad with delight, recklessly rolling into the deepest snow-drifts he could find.

We remained on shore three days, and then moving up in a boat we walked over the tundra till we found more reindeer, with which we went to Okshin on the Petchora. The journey was full of incidents. Among other things, I may mention that I found my fame as a doctor had spread, I suppose because I had dosed them with ginger and pills on Kolguev. One of my cases was that of a poor old man who had badly cut his knee with an axe. And after making him a poultice I turned to go, when there came a voice behind me, 'This is for him who came to us in our trouble, this is for our friend,' with which pretty speech his son offered me a blue fox's skin.

But I set out with the object of telling you about Kolguev Island, and had certainly no idea it would take so long in the telling.

You will have had enough of us and our adventures by this time, if indeed your patience has carried you so far. We travelled from Okshin up the river to Ust Tsilma, from there overland to Archangel, from there to Vologda by sleigh, thence to Moscow, whence I sent Hyland home by Riga, going myself to St. Petersburg, where the Tsar's body was lying in state, and so to England.

I arrived at Queenborough on November 19th.

To tell you the details of this journey would take a separate book by itself, for it opens a new chapter on mainland Samoyeds and on Russian peasant home-life.

If we had chosen to wait for the winter we should have had no further difficulty in crossing the frozen rivers and the snowy tundra than to sit on a sleigh, as others have done before. But, if Hyland was to be home for his Christmas customers (a serious consideration in his little trade), it was very necessary to start at once. And so it was that we fell on a time when not a reasonable soul ever took that journey, because the rivers were half-frozen and very dangerous, and the swamps impassable, as was supposed. So that even the Government service stops during this season of 'Rasputa,' and all contracts are off for a month, and it is a time of holiday, or, as the Russians say, of 'Stroke.'

But I have written enough for this time. If this book should win as much interest as may serve to warrant it, perhaps I may yet write the story of our travels from the Kolokolkova gulf to old England itself.

How we lived with the natives, and the friends we made, how we attended a peasant wedding, and of the quaint ceremonies that followed on it, how we crossed the Petchora and took the horses over the half-frozen Tsilma river, how we wandered in the forests, and how we slept at nights, how we crossed a deep stream on trees which were thrown from bank to ice, how the wolves came and carried off a pony, how a midnight row came about on the ice of the Volga—all these and many other incidents would be there particularly described.

And now I take my leave. My honest companion, Thomas Hyland, is once more at home and happily united to the girl of his choice, with my very best wishes for his future success. It would take more than ten thousand reindeer, I think, to get him again to Kolguev.

Old Sailor, too, has done with travelling. Happy and safe in an English kennel, he is doubtless proving a mine of yarns to his poor untravelled companions. Sometimes when he lies asleep he will yap and growl amazingly, while curious twitchings take him. And then I think he is holding his own with the wolf-like dogs of Kolguev.

It is May in England as I write these closing words. It would be hard to conceive a greater contrast than that of Kolguev (with its fog-swept wastes, its wild life—half a freedom and half a shrinking from the fingers of the far-reaching iron hand) and such a scene as lies before me now.

Through the open window is borne a scent which has no name. Not from briar and violet alone it comes, but from daisy, from willow-catkin, from the very turf of the growing grass. With every passing shower it changes, and with every touch of the sun.

Under a cloud of gnats the trout are rising, till the pool below the cattle ford twinkles and glints again.

Further yet—where the beech wood cuts a line of emerald across the hill—the hollow is billowed in blossoms pink and white, and a jet of blue smoke feels its way up among the trees. And faintly across the meadows comes the barking of a collie; for the men are folding sheep.

All this—the scent of May, and the softness, the homestead and the quiet rhythm of its life—seems to me the expression of a certain rest and confidence that belong to the beauty of strength. I have wandered among many people, but I have not elsewhere found exactly this.

'Dear old England!' is a phrase which comes very glibly to the lips. But to mean it I think you must be content for a spell entirely to sever yourself from the land that is laid as a garden—the land where Victoria is Queen.

It is worth doing, because it teaches you, as it only can, the perfect meaning of the old burden, 'Home, sweet Home.'

THE END.

PART VII

An Abstract from the *Saxon's* log—Remarks on the Samoyeds of Kolguev—Note on the Samoyed language—Note on the geology—Note on the flora of Kolguev—List of flowers — Note on the birds recorded — List of birds—Note on the mammals—Remarks on the climatic conditions

FROM THE *SAXON'S* LOG

By M. W. POWYS

TO SHOW WHY THE *SAXON* DID NOT TAKE US OFF THE ISLAND

June 22nd.—Lovely morning. Sailed away from Kolguev at 1.30 A.M.; our course west. At 4 A.M. steaming half-speed, with heavy ice-pack to the north; much clearer to the south. At 10.30 A.M. began to get clear of ice, of which we lost all sight by noon. From 4 to 4.30 P.M. were passing heavy pack ice to the south of us.

23rd.—Noon, lat. 70° 3′. Sea heavy towards night. At 9.30 P.M. sighted land on our port-bow. A lovely day.

24th.—Anchored in Vardö harbour, 7 A.M.

27th.—Left Vardö 8.30 P.M., shaping course for South Goose Cape.

28th.—Dull morning. Fine afternoon. Heavy sea behind us.

29th.—Dull misty morning. Much colder. Position at noon 71° 4′ N., 45° 48′ E. At 2.30 P.M. (our course then E.NE.) came on loose ice right ahead, and stretching to the north. We went through this for about an hour; then, getting clear of it, set our course again E.NE.

30th.—Dense fog from 1.30 A.M. to noon. Heavy ice to the north. This ice, which seems closely packed to the east, we were dodging all day.

July 1st.—With a dense fog the ice was as heavy as ever. About noon we distinctly made out land (lat. 72° 1′) from which we were separated by closely-packed ice. Abandoned our northerly course and turned southwards. Heavy swell all day.

2nd.—Foggy. Dodged about the ice all day.

3rd.—The weather a little clearer. We were able to stand in nearer the land, but could not reach it.

4th.—Foggy. We followed the line of the ice; but it took us west. As it was impossible to reach the land I decided to put about; so, getting clear of the ice at 6 P.M., we shaped a course for Sviatoi Nos (Kanin).

5*th.*—Heavy sea. Screw out of order. Reduced to sails.

6*th.*—Sea less towards evening. Got up a little steam.

7*th.*—Anchored off Solombola (Archangel), and made arrangements for repairs.

8*th to* 23*rd.*—Detained during progress of repairs.

24*th.*—Left Solombola for Kolguev.

26*th.*—Foggy. Sighted Kolguev at 1.30 P.M. Tried to land at the Gusina, but a heavy surf made it impossible. Lay a course again for Novaya Zemblya.

27*th.*—At 3.30 P.M. came to anchor in five fathoms off Meshdoshapsk Island.[1] At south-eastern corner of this island are two crosses—very conspicuous. At 8.30 we got up steam, and had out the dredge, taking sea-weeds, shrimps, and small starfish.

28*th.*—Anchored in five and a half fathoms some eight or nine miles further up the coast. Here we landed and went to inspect a house that I had noticed from the ship. We found it in good order, with windows of glass. Inside were the remains of a white fox, and outside we found similar remains and a great quantity of reindeer horns. We left in afternoon. The soundings were very uneven. Entered Nechvateva bay on the mainland at 4 P.M. Here we ran on a rock, but, getting off again, let go anchor in nine fathoms, 200 yards off shore, and landed. I saw reindeer tracks and ducks (very wild). The country, which was very hilly and stony, contained several small lakes.

29*th.*—Pulled about four miles up the bay, and then landed and walked inland. Saw numerous tracks of deer. This was a country of hills and valleys; the former stony and rough, the latter swampy and filled with lakes. Ducks were numerous, but they and the geese, of which we saw a few, were very wild.

30*th.*—Returned to the island. Saw six deer which I followed, but unsuccessfully.

[1] 'Meshdoshapsk Island is very swampy, with low hills in the middle. It is filled with lakes. In the northern part deer are numerous. Anchorage good. The drinking water here is good. Careful navigation is most important. In the Kosten Straits the bottom is very uneven. The entrance to Nechvateva bay is dangerous.'

31*st*.—Moved up northwards for some six or seven miles, and then, sighting nine deer, landed. I secured one.

August 1*st*.—Blowing hard. Had hoped to get away, but the weather was too bad.

2*nd* and 3*rd*.—Still blowing hard. Remained at our anchorage.

4*th*.—Wind moderated. Got away for Kolguev early, and at 9 P.M. sighted land.

5*th*.—Anchored at 4 A.M. somewhere it seemed off the Baroshika river, and about four or five miles out in six fathoms. Later on moved in nearer till the soundings showed four fathoms. Then I went off in the cutter to try and land, but could not, owing to the sand-banks, which were separated from the coast by a stretch of open water. During the day we cruised up and down trying to find an entrance. As we could not find one, I returned to the ship and we again followed along the sand-banks. At 10.30 P.M. ice was reported to seaward, drifting in towards land. We edged nearer in, proceeding very cautiously, as the soundings rapidly lessened from eight to six fathoms of water. At last we were compelled to turn back because the ice began setting towards the land. 'If we go *on* we shall be wedged between ice and sand-banks; if to *seaward*, get stuck in the ice; if to *landward*, on a sand-bank.'

6*th*.—Having retraced our way we rounded the north of the island, and anchored off the Gusina river at 10 A.M. From then up to 2 A.M. ice was constantly reported to seaward. I went ashore and found two notes buried there in a tin by Trevor-Battye. In one of these he said that he and Hyland, having found the Samoyeds, are going across to Scharok harbour. We believed this to be the point from which we were driven away the night before by the incoming ice. With the wind E. by N. it would now probably be blocked with ice, and therefore impossible to enter.

I therefore decided to go back to Vardö.

7*th* and 8*th*.—Dense fog nearly all the time. Got into Vardö on the morning of the 8th.

9*th* and 10*th*.—Blowing hard.

13*th*.—Tromsö.

21*st*.—Peterhead.

REMARKS ON THE SAMOYEDS OF KOLGUEV

THERE are at the present moment upon Kolguev fifty-nine Samoyeds. Of males, twenty-four, of the opposite sex, twenty-six. These include children, but children old enough to work. The remaining nine are babies, or children of helpless age.

Most of these people—all I think but five—were born on the island. But I was unable to determine at what period in their family history their connection with Kolguev began. Some of them had Kolguev traditions which seemed to reach far back. But all referred their relationships to one or other of the mainland divisions of their race. Thus Uano claimed to be a Timanski, Marrk a Kaninski Samoyed.

On the Timanski tundra one finds that the choom is the unit of Samoyed life. Only abnormally are chooms grouped together, *e.g.* near a village or town. In this case the Samoyeds are either beggars or are labourers to the Russians. Temporarily on the tundra chooms may be pitched in close relationship, as the owners pass in travelling, or are drawn by some common object—a seal hunt, for example. But each family by itself is the principle of existence, this or two together; but then one will hold the parents, one the married eldest son.

And on Kolguev it is just the same. For the purposes of life and reindeer pasture the island is by mutual consent divided into districts, as shown by the names on my sketch-map. Thus Uano has the Lower Pesanka, On Tipa the Upper Pesanka (and, speaking roughly, the north-wester part of Kolguev), Marrk the Gobista—and so on. They scrupulously respect each others' rights.

A father and his eldest son had on Kolguev their reindeer in common, or rather managed together the family herd, for the deer belonged strictly to the father until his death, when they would pass to the eldest son.

When a younger son desires to marry, his father gives him a certain number of reindeer. This number is not constant. It is the result of bargaining between the parents; and is determined by such considerations as the advantages of the match, by the amount of property which has previously been asked to the family by this son's hunting, and by other such points. The girl also brings with her a dowry. In the case of Uano's daughter, Ustynia, it was, as I have said, five young bulls.

All Samoyeds are compulsorily affiliated to the Greek Catholic Church. The priest quoted before (*v. supra*) has told us how he went to Kolguev to baptize. But it is many years now since any priest was there. None the less the yearly visits of the Russian traders (who, though they demoralise these poor people with vodki, yet put them to religious exercises), added to the vague sense of a compelling, though distant, control in the Archangel Government, serve to keep alive in them a thin thread of the Christian idea.

Possibly because of the isolation of their home the Kolguev Samoyeds may lean to their own old faith more than those of the mainland, though this I do not know. On Kolguev, at any rate, there is but one family, that of On Tipa, who consistently keep the *ikona* on view in the choom. The remainder carry the bolvans—or, as they call them *shya-dèy*—which represent the God Nûm or Philibyambierchi.

I remember to have read somewhere of a ceremony followed when carrying the dead man out of the choom, but the Samoyeds' traditional treatment of their dying, according to Uano, was that of laying the dying man on the moss of the tundra. He was not abandoned till dead. Only towards his last gasp he was brought out and laid there. As soon as he was dead he was stretched straight out on his back with his hands at his side, the attitude in which these people sleep. But now the dead are buried, and although the broken sleigh and certain domestic utensils of the dead man are still taken to the Holy Hill of Nûm, yet a cross is set up over his grave. So is there a kind of dual regard. Nordenskiöld suggests that this practice has reference to the future needs of the dead man. But, closely as I questioned the Samoyeds, I could not extract from them any explanation but this, namely, that the sleigh was broken to show the man was dead; though it is true that Alexander's explanation was different. He said it meant that the man was now 'like a broken sleigh.'

On the mainland, if, as I believe, I am credibly informed, it is the custom for a Samoyed to take a girl as his wife on trial for a year. If at the end of that period he is satisfied with her, the compact is ratified, either by marriage by a priest or according to native uses. If not, then the two persons are free to marry again; and if a child has meantime been born the father is bound to maintain it. On Kolguev there is no priest, and I do not think that this practice of a trial association obtains. I believe that when a couple are once betrothed they remain constant to one another.

One not infrequently sees half-breeds on the mainland, but of the conditions of Samoyed moral relationships there *inter se* I know nothing. But in the case of so small a community as that on Kolguev it is obvious that, in this respect, it must strictly protect itself. And this it does.

The Samoyeds are prisoners on their island. They have no boats which could venture across that fifty miles of open sea. In the old days, when there was traffic with Mezen and Indiga, they may have passed more frequently; now it is only occasionally that one is taken across. Last year but one returned, and he had only crossed with Alexander the Russian the summer before.

The Samoyeds of Kolguev suffer from none of those complaints which, introduced by the young Russian soldiers, sometimes disfigure their people on the mainland. Every individual on the island appeared to be sound and healthy, with the single exception of young Mekolka, who had some lung trouble.

I was very much struck by the fact that the Samoyeds appear to be able to live entirely without vegetable food, if we except the summer berries (which they do not, however, preserve). I understand that the Eskimo commonly eat the contents of the deer's paunches. Our Samoyed never did this; they always threw it away.

Family affection among the Samoyeds is very strongly developed. It would be impossible to find greater evidence of this among any people. Another extremely marked character among them is family order. All every-day offices and occupations are carried out by a well-defined method and sub-division of labour. I never saw a single instance of anything approaching a family quarrel. And I never on any occasion knew a Samoyed man or woman lose their temper. My companion, Thomas Hyland, often remarked to me what a fine example they would be in this respect to many of us in England. I could not help contrasting them in these ways with my old friends, the Indians of North America, greatly, I fear, to the Indians' disadvantage.

May I venture to add a few reflections called up by the subject?

I have heard it said that the Samoyeds as a race are useless to the state and to employers of labour. It is said of them by some, as of the Indians, 'they will not labour and cannot be taught.' I venture to believe this is a mistake. To try and make agriculturists of a pastoral and nomadic people is a contradiction in experiment foredoomed to failure. And I am sure that that no attempt on the part of the government to take them away from their associations, and to make them work as soldiers or sailors under masters who do not understand their character, would be more successful. One bar to this, if no other, would be fatal. I refer to the question of language. Very few of the Samoyeds can speak the Russian language proficiently. Always they mix up with it words of their own, or corrupt it to a pronunciation easier to themselves. And though there results a pseudo-tongue, intelligible enough to the northern traders, I doubt if any Russian could understand them at the first contact.

But they are very handy sailors (if one may argue from those with whom we crossed, and those we saw on the Petchora), patient and successful hunters and fishermen, and admirable workmen with such tools as they understand. No man can repair a damaged boat more quickly than a Samoyed, and from the roughest drift-wood (which an English carpenter would throw on the fire) they fashion bows, arrows, sleighs, spoons, drinking-cups, bullet-moulds, and a variety of articles of every-day use. I do not think that the Russian Government realises that they are not a worthless people. The few traders who regularly employ them appreciate them more fairly. They know that in the Samoyeds they have not only men who will bring them skins, but reliable workmen, who go on at their barrel-making or boat-building, or what not, regularly and steadily, year in and year out, when vodki is kept from them. Alexander Samarokoff assured me of this, and he is a large employer.

In view of the Trans-Siberian railway, the Russian Government has in the Samoyeds of its tundra an agency which, if allowed, will do no small part in developing the resources of Arctic and Northern Russia. As workmen they might be useful; as carriers they would be pre-eminently so.

But one consideration demands imperative attention. The great failing of the Samoyed—one which he shares with the peasants of Northern Russia—is a love of vodki. Unless the Russian Government will make it a criminal offence to give spirits to a Samoyed, the Samoyeds are doomed.

At the present moment the Samoyeds are I believe responsible, for the purposes of taxation, to two overseers. Some extension of the system with a view to supervision might be well. By men who understood their characteristics, and who would treat them kindly, they could easily be managed and become willing workmen. Anyhow, it will be the wisdom of the Russians to nurse their Samoyeds and not to kill them. And as trade advances these most useful and interesting people, unprotected, will surely die.

NOTE ON THE SAMOYED LANGUAGE

THOSE who are inclined to make a study of the Samoyed language will go of course to the pages of Castrén. It would be absurd for any one but a philologist to attempt a serious consideration of it.

But it has seemed to me that just a few hints from one who spent three months in trying hard to speak with the Samoyeds, may be of some practical use to any other naturalist who shall go among those people.

The first point, of course, is to get hold of the proper names. When you once get these, then you can string them together with Russian, and this is easy enough to learn to speak. I do not mean good Russian (and they do not always understand that), but *bad* Russian, the Russian of the peasants of the North. And the Samoyeds, even among themselves, habitually use a good many corrupted Russian words. For example, in *mahlsa* (a coat), one sees the Russian *malitsa*; in *soyuk*, the Russian *sovik*.

Only bit by bit can one hope to put together a connected sentence in an agglutinative language. And although I certainly did succeed in understanding and in speaking a little, I am quite sure that I could never write it down in an intelligible form. The Samoyed tongue is largely breathings and inflections, which cannot be expressed in letters.

In the words which follow the vowels are, of course, as in Italian. Perhaps this key will help :—

a=ar	
i=ee	*ei*=i.
e=a	*rr*=r *rolled.*
ey=a	
er=air	
ü=French u	*h* final is like h in *loch.*

The pronouns are puzzling. Castrén, I believe, gives a large number. I never could detect more than three :—

I and mine, *mein.*

You and your, *puddrr* (with a point of the finger at the person indicated).

He and his, *puddrr* (without the point).

It will thus be seen that just as the Samoyeds use no ordinal numbers, so they use no possessive pronouns (properly so called). Instead of this, the personal pronoun, when used as the possessive, has the function in some, if not in all, cases of changing the word it agrees with. Thus, 'sleigh' is *hàn*, but 'my sleigh' is *mein hanòo*; 'knife' is *har*, but 'my knife' is *mein harvòo*; 'son' is *neu*, but 'my son' is *mein neydòo*.

I am aware that a list of Samoyed words is given in Rae's *Land of the North Wind*. The reader can compare my list with his. I have deliberately forborne from looking at that list myself, because I did not want to confuse my recollections of sound. It is very possible that other writers have been more etymologically correct, but this I can promise my naturalist friend, that if he can pronounce the words as I think my spelling will enable him to do, then the Samoyeds will understand him; and this in a language where a shade of difference means sense or nonsense. Of course there are different dialects, but the words I learnt on Kolguev were always understood by the Samoyeds I met with west from the Petchora.

SOME COMMON SAMOYED WORDS FOR FUTURE TRAVELLERS.

ANIMALS.

Walrus, *diùlieh*.
Polar bear, *hèvdü*.
Fox (red), *tchuòrnia*.
Fox (blue), *nohò*.
Dog, *wenquòh*.
Blue-bottle fly, *dorndàftsa*.
Reindeer fly, *pìliüh*.
Musquito, *nianìnk*.
Seal, *siùk*.
Grey seal, *zèi-itz*.

FOOD.

Soup, *yah*.
Salt, *sìrr*.
Bread, *rìèska*.
Flour *yìak*.
Dough, *goòsah*.
Goose-fat, *yùbtaw yoorr*.
Bread made with fat, *am-du-dah*.

ADJECTIVES.

Light, *sìvichin.*
Heavy, *sangoòa.*
Hot, *ibka.*
Warm, *hiyèrpi.*
Cold, *di-aìt-si*; *tìchi.*
Wet, *schù-vir*; *sa-nòo-y.*
Dry, *ha-sòo-y.*
Much, *doo-tàm.*
Little (not much), *ork-hòh.*
White, *fèrr-koh.*
Black, *barìdia.*
Good, *sow-wàw.*
Bad, *we-o-dü.*
Light (not dark), *yàrr-li.*
Dark, *limbia.*
Hungry, *or-mùn-ga-wum.*
Big, *arrk-ha.*
Little (not big), *nù-dya*, *ni-oìt-sko.*

PREPOSITIONS.

After, *poò-noh.*
Before, *ni-èr-nia.*

NATURAL OBJECTS.

Grass, *sti-èl-ka.*
Flower, *om-doh.*
Green-moss, *tow-qwoh.*
Reindeer moss, *ni-ar-dèy.*
Bog moss, *ni-àrt-si.*
Wind, *mi-èrrt-sha.*
Rain, *sàr-rù.*
Ice, *sàrlba.*
Snow, *hardt.*
Frost, *hein.*
Hail, *sièrrts-a.*
Light, *yàrr-li.*
Dark and night, *lìmbia.*
Thunder, *moòn-ta.*
Rainbow, *su-òi-vrr.*
Dew, *yùb-la.*
River, *ya-ha.*
River source, *ya-ha màh.*
Mountain, *hoy.*
Island, *mar-ra-òh*; *awh.*
Cape, *zàb-li.*
Sea, *sièrr-koh*; *yamh.*
Lake, *toh.*
Fog, *suniàw.*
Sun, *hei-yàrr.*
Moon, *iri.*
Water, *yi.*
Fire, *too* or *tuh.*
Stone, *peh.*
Bird, *ti-èrr-ti-koh.*
Fish, *hàrr-lih.*
Cloudberry, *mà-ra-ga.*
Cranberry, *yez-dey.*

TIME AND THE SEASONS.

The Months—
August, *Si-èrltz.*
September, *Too-ròitz*
October, *Mal-coòmps.*
November, *Pei-o-dèy.*
December, *Sidia Pei-o-dèy.*

January, *Limbia.*
February, *Yàrrya.*
March, *Si-iss-nit.*
April, *Nun-nèi-i-nitz.*
May, *Na-witch-y iri.*
June, *Màhng-tis iri.*
July, *Yab-toòz iri.*
Spring, *ù-nòy.*
Summer, *tùng-ùr.*
Autumn, *ùr-ri-òy.*
Winter, *sir-rèy.*

A month, *iri* or *yiri.*
Night, *limbia.*
Day, *yarr-li.*
Morning, *sien-nìtz.*
Evening, *pèi-oo-shìm-bü.*
To-day, *to-qui-à-lia.*
Yesterday, *ü-yàrr-li.*
To-morrow, *hòo-in-yàn.*
Day after to-morrow, *hòo-in-nı-tyàn.*

NUMERALS.

1, *apoi.*
2, *sìdia.*
3, *ny-àrr.*
4, *tierrt.*
5, *tiet.*
6, *sùm-la.*
7, *mot.*
8, *sì-ùgh.*
9, *sìn diet.*
10, *ha-bi-èg-you.*
11, *òr-bi-ò-go-nu.*
12, *sìdi-ègonu.*
13, *niàr-ri-ègonu.*
14, *tierrt-egonu.*
20, *you.*
21, *apoi-you.*
22, *sidi-you.*
100, *yōūrr.*
1000, *you-òr-na.*

DOMESTIC ARTICLES, ETC.

Knife, *har.*
Cooking-pot, *yud.*
Snuff, *hei-nìb-chur.*
Buckle, *nỳ-èss.*
Axe, *toov-ka.*
Nail, *yès-dìep*
Ring, *o-di-è-sa.*
Net, *hei-lim-boòr-ga.*
Dog-lead, *sàr-ròw.*
Bow, *urn.*
Boat, *ar-nòh.*
Oars, *là-vy-èi.*
Bighone, *si-lìtz-sim-pa.*
Small hone, *si-èrltz.*
Case for hone, *si-ì-zia.*
Skates or skiddor, *lum-bàh.*
Hammer, *mul-to-dà-da.*
Fox-trap, *yes-ièn-go.*
Tow, *yern.*
Rust, *pòo-rom-a.*
Gun, *toò-nya.*
Horn, *niàmd.*
Steelyard, *ye-so-lò-ka.*
Bread spatula, *si-ù-quòh.*
Float for net, *po-lòw-a-ka.*

Dress.

Girdle, *nì-e-di.*
Child's cap, *schwòy-ut-sa.*
Girl's dress or panitsa, *panü.*
Men's coat or malitsa, *zarka.*
Cotton covering to malitsa, *zarka imboot.*
Inner cotton shirt, *àh-yi imboot.*
Sealskin boots (girl's), *pi-où.*
Sealskin boots (men's), *pimüh.*
Gloves, *obah.*
Collar (fur), *yì-kirr.*

Parts of Body.

Hand, *wud-dóh.*
Finger, *om̃-bya.*
Thumb, *pìk-chya.*
Palm, *pya.*
Neck, *yeek.*
Nose, *pu-i òw.*
Eyebrow, *neèr-chèen.*
Mouth, *ny-àm.*
Teeth, *tì-bya.*
Hair, *erpt.*
Beard, *moon-sa.*

Sense and Emotion.

Sight, *yàrr-ka.*
Tears and sorrow, *yarrn-ga.*
Blindness, *sèy-yòw.*
Laughter, *pì-sing-a.*
Head and memory, *tei.*
Whistling, *shu-ga-ring-on.*
Singing and song, *huntz.*
Pins and needles, *tì-et-si.*
Dancing, *ni-èrr-nia.*
Jumping, *nòw-o-toò.*

Relationships.

Father, *nìzia.*
Son, *new.*
Wife, *nì-ow̃, poo-hoò.*
Husband, *wa-sìl.*
Baby, *adski.*
Samoyed man, *nya-nìtz.*
Samoyed woman, *nya.*

Miscellaneous.

Hurry! *oò-tan-hèy!*
Start sleigh, *hèi-hàn!*
To the left, *sìet màrquoh.*
To the right, *màn-yi-àrquoh.*
Give me, *tàlnya dahdt.*
Tracks, *oòdh.*
I am come (on foot), *mein miñgùm.*
I am come (on sleigh), *mein doo-tàm.*
I know, *mein ten-yoo-wàm.*
I do not know, *mein ye-ha-rhàm.*

NOTE ON THE GEOLOGY OF KOLGUEV

KOLGUEV ISLAND lies between 68° 43′ and 69° 30′ N. lat., and reaches from long. 48° 15′ to 49° 55′ East of Greenwich. Its greatest length from north to south is about fifty geographical miles, its extreme width from west to east is about forty geographical miles. Its distance from Sviatoi Nos, on the mainland of Arctic Russia, is fifty miles. The soundings between the island and the mainland are under thirty fathoms. Though in a broad sense we might speak of Kolguev as an extension of the continental tundra into the Arctic Sea, yet a critical examination of the surface geology of the island clearly shows that the modern Kolguev has not been connected with the mainland since its elevation above the sea. The surface geology of Kolguev plainly tells us the story of recent upheaval, so striking a feature in most Arctic and Polar lands, and which geologists, who have made those regions of the earth a study, are constantly impressing.

During my three months' stay in Kolguev, in my many journeyings and careful examination of river-beds and sections, both in the ravines of the more elevated high-lands, and along the steep mud cliffs that so generally fringe the sea-coast, I came on no rock surface *in situ*, nor on any section where there was an exposure of stratified rock. I may at once say that I am completely in the dark, as far as my observations go, in regard to the solid geology of the island.

The superficial area of the island is sharply divisible into two portions. Speaking generally, the northern two-thirds are high ground, which consist of peat-covered or of bare ridges intersected by gullies, and enclosing small lakes and swamps, and the remaining portion to the south is a dead flat of grass, bog and peat-levels reaching to the sea.

Of mountains there are on Kolguev none that fairly deserve that name. Sowandeyi and Siecherhur, the two highest points on Kolguev (named by the Samoyeds from hills of the same name on the Timanski tundra), are, as near as I could estimate them by aneroid, two hundred and fifty feet above the level of the sea. They are essentially great heaps of sand, terminating in peaks and ridges, carved into a curious confusion of crater-like hollows and gullies by the melting snows. On these hills the winter snow remained longest, and had not all melted there when I visited them on August 26. Siecherhur, a 'bolvan mountain,' is really a remarkable hill in appearance. Thickly covered with grasses, almost completely circular at its base, it rises suddenly from the general level—a conical hill, as smoothly and regularly formed as though an old artificial camp. Another hill, Honorohur of the Samoyeds, is a very conspicuous object from far round, and very characteristic in shape; so that it is called by the Russians 'Lodka' (the boat), for that is the appearance it presents to them when seen from the mouth of the Pugrinoy.

There is not the slightest evidence to suggest that Kolguev ever supported an ice-sheet. There is not, over the entire island, a trace of a moraine, either lateral or terminal. I came across no beds of true till, or typical boulder-clay. All the deposits are referable to wave and oceanic forces, and Kolguev is essentially a water-and-ice formed island.

It is not within my province to suggest from what river system the material was derived which forms the modern island of Kolguev, whether it was brought down by the Petchora or by other rivers, but there can be no doubt that it represents the wearing down of the adjacent continent, and the strewing of the sea bottom with ice-borne erratics.

No better evidence to the formation of Kolguev could be obtained than that afforded by such a point as the summit of Mount Sowandeyi. This is indeed a key to the whole geological history of the island.

He who stands there sees before him to the north a wide high plain of peat and lakes, studded with hills which have the appearance of gigantic ant-heaps; a nearer examination of these shows them to be formed of sand, clay, and *débris* of small stones. They exactly correspond with the constitution of the hill on which he now stands.

This hill has the shape of a broken and hollow cone grass-clad on three sides. The traveller who ascends to the top finds himself suddenly at the edge of a funnel-shaped crater, corresponding to the three outer sides of the hill. The walls of this crater are entirely formed of sand and clay, in which no horizontal stratification can be traced, for all has been washed by snow and water from above downwards into streaks and runnels that lead to the foot. Here is collected a confusion of large stones evidently washed out of the bed in which they lay. The broken fourth side of Sowandeyi is open to the north, and by that exit passes the water which streams from the meltings.

This double fact of peat-levels and of sandy or clay mounds and ridges finds its exact counterpart on the eastern coast, and in the results of the action of the tides to-day.

In the first place, you find the wide peat-levels reaching inland fringed on the coast-line with a solid higher rim of sand and clay. And secondly, the process by which these were formed is again repeated tide by tide and month by month outside.

There is the harbour mud, partly brought down by the streams and partly the result of disintegration of the clay banks by frost, lying—a wide lagoon of mud—within the outer banks of sand. To the south this mud passes insensibly into the peat-levels by a traceable process. First, the mud or ooze, then this growing firmer, filled with roots and covered with grass, and this again giving place to moss and lichen, till the 'peat-hag' itself is formed. Doubtless this rising is partly due to superimposition of material, but to a far greater extent it is to be attributed to the actual rising of the whole area, in agreement with a condition known to be true of the tundra and of Scandinavia to-day.

The outer sand-banks of which I have spoken (which exactly correspond in character with the sand we examined from the soundings outside) are being periodically piled higher by the action of the waves. So rapid is this action round Kolguev Island that the Russians told me that, in the thirty-five years during which they had visited the island, the general relationships of one channel to the other had completely changed.

The time will come when all this region, now under the influence of the tides, will be added (a new peat level and a new inland beach) to the area of the island.

Surely therefore, this—taken in connection with the actual geological conditions of the interior of the island—is a sufficient index to the growth of Kolguev as a whole.

If the question should further be asked, 'How, and from where, then, has this *débris* come?' I think for answer one may safely say as much as this:—partly by some great river system, and partly by the ice. The granite, sandstone and limestone (?) boulders found on Kolguev are almost certainly ice-borne erratics from the mainland tundra and Novaya Zemblya. Such a striking sandstone as that figured here, which, lying now between the snow-banks, shows round its side striations made by ice, may safely have come from the tundra, where no traveller can fail to be struck with the sand-rocks which take there similar eccentric shapes from the wearing of such forces long ago. But at least some fossils in my collection taken from boulders on Kolguev seem to point to a more distant source as kindly described for me by Mr. Ethridge. They belong to the Upper Silurian period. Two are corals. *Halysites catenularius* and *Cyathophyllum truncatum.* One is a gasteropod, probably *Naticopsis.*

Note.—I have ventured to incorporate with this part of a paper read before the Royal Geographical Society, an abstract of which appeared in *The Geographical Journal* for February 1895.

A NOTE ON THE FLORA OF KOLGUEV

A GLANCE at the remarks on the climate of this island will show how very severe the conditions are under which plant-life there exists. As a consequence of the frosts of June, the frequent cold-driving mists, the prevalence of northerly gales, and the exposed character of the country, plants, in a majority of instances, tend to a dwarfed condition and an abnormal habit of growth. Thus we find the common yarrow (*Achillea millefolium*) reduced to a plant some three inches high and tomentose. The woolly willow (*Salix lanata*), which on the mainland, not ten miles from the sea, is as high as a man, grows nowhere on Kolguev higher than the knee, and only reaches that height in the most favourable spots.

Plants, again, are exceedingly late in flowering and in ripening fruit. We left the twin-flowered violet (*Viola biflora*) in full flower at Tromsö on June 11. Yet on Kolguev it had only just come into bloom on August 2. The cloudberry (*Rubus chamæmorus*) was not generally ripe with us till August 25. Yet the Russians told me that these berries were 'over' on the mainland when they left the Petchora on August 11. We saw no fruit of *R. arcticus*, and the Samoyeds all agreed in saying that it bore none on Kolguev.

Of the ninety-five flowering plants found by myself on Kolguev, some sixty-three are recorded as British. But of these, as one would have expected, many are in Britain either rare or exceedingly local. Thus *Arabis Alpina* has only been recorded from Skye;[1] *Draba rupestris* is 'rare on some of the higher mountain summits of Scotland and north-west Ireland.'[2]

Of the plants which are not British some have nevertheless a wide palæarctic and some a circumpolar range. To the former class belong,

[1] H. C. Hart., *Journ. of Bot.*, 1887. [2] Bentham and Hooker.

for examples, *Delphinium elatum* and *Dianthus superbus* (which reaches through Arctic Europe and Asia to Japan); to the latter class, among others, *Valeriana capitata* and *Primula stricta.*

Although, as was to have been expected in the case of an island only fifty miles from the coast of Arctic Europe, the floral affinities of Kolguev are with the mainland, yet there are some exceptions, which seem to me worth noticing.

I am glad to take this opportunity of thanking the following members of the British Museum Staff for careful comparison of my plants with the type-forms in that collection, and for much trouble in identification—Mr. James Britten, Mr. Antony Gepp, Mr. E. G. Baker, and Mr. A. B. Rendle.

I find three instances of boreal plants which appear not to have been recorded from Arctic Russia—*Stellaria Edwardsii*, *S. humifusa*, and *Antennaria carpathica.* Both of these are said to be found on Spitzbergen and Novaya Zemblya.[1] Finally, the apparent absence from Kolguev of such well-known northern forms as *Saxifraga oppositifolia*, *Mertensia maritima*, and the beautiful azalea-like *Ledum palustre* (*L. dilatatum* Ait.), so striking a feature of the tundra, is at least remarkable.

Since my return to this country I have been carefully through the list of plants collected in Kolguev by Dr. Ruprecht. They will be found in a work by him entitled *Flores Samoyedorum cisuralensium.*[2] The work he accomplished during his short visit to Kolguev was admirable. He collected flowering plants only. It was inevitable that he should miss several of the plants which I have in my own collection, but, on the other hand, he records several which I did not see. Dr. Ruprecht defined as species certain plants which are now only regarded as climatic varieties (*e.g.* he described and figured as new species no less than seven forms of *Poa*). But of those which I believe to be accepted as species, and which do not appear in my list, I give the following, as a tribute to his work, and with the idea of making the Kolguev list more complete, and therefore more useful to others:—

Ranunculus lapponicus L.
R. Samoyedorum Rupr.
Aconitum septentrionale Willd.

[1] *Cf.* Nyman's *Conspectus.*
[2] *Beitrage zur Pflangenkunde der russischen Reiches.* St. Petersburg, 1845.

Eutrema Edwardsii R. Br.
Draba muricella Wahl.
D. hirta L.
Viola epipsila Ledeb.

(Of this he says, 'Very rare, and all growth of leaves is wanting.')

Parnassia obtusiflora Rupr.

('Allied to P. grœnlandica.')

Sagina intermedia Rupr.
Cerastium arvense L.
Potentilla Comarum Nestl.
Epilobium lineare L.
Saxifraga foliolosa L.
S. hieracifolia L.
Pachypleurum alpinum Ledeb.
Adoxa Moschalellina L.
Senecio campestris DC.
Vaccinium Myrtillus L.
Pyrola uniflora L.
Androsace septentrionalis L.
Juniperus communis L.

('Nor is it absent from Kolguev, where it is to be referred to *J. nana* W.')

POLYPETALÆ

RANUNCULACEÆ

Ranunculus pygmæus Wahl. Pigmy Buttercup.

The pigmy buttercup was very common on the east side of Kolguev, on damp clay banks, and on the edge of the mud-flats. It was not in full flower till September.

R. acris L. Var. *pumilus* Hartm. Meadow Buttercup.

This dwarf form of one of our commonest English buttercups was one of the earliest flowers on Kolguev. It flourished best on the sunny side of the gullies among thick grass. I first saw it on our second landing on June 17 (on a bank on which the snow was still lying in patches), in company with *Oxytropus* and *Myosotis*. When in full bloom the heads form a most brilliant cluster, on account of the low growth of the plant.

R. Pallasii Schlecht. Var. *β. minor.* Rupr.

I have referred earlier to my first meeting with this plant on Kolguev. It presented characters so distinct from any Ranunculus I had before seen that I took the greatest possible interest in it. I brought down a quantity of it from the Pesanka and from the Swan Lake, and kept it alive by our Scharok waterhole till we came away. Eventually I sent the plants to the Botanical Gardens in Oxford.

The petals (6 or 8), are white—not semi-transparent white as those of *R. aquatilis*, but white as *Parnassia palustris*. The sepals are 3 or rarely 4.

The flower is very sweetly scented—a point which seems to have been quite omitted in previous descriptions. The plants I brought from Kolguev exactly agree with the description of Ruprecht's *β. minor*. It is smaller than the type, and the cauline leaves are all unlobed. This plant grew only in the wettest *moss* bogs, in which its stems ramified, rooting at the joints.

Mr. Baker, to whom I am indebted for identification of this plant, has kindly supplied me with the following notes :—

The type, as figured by Schlechtendal in his 'Animadversiones,' is large and fistular, and with generally lobed leaves. I have examined the specimens both at the Natural History Museum and at Kew, and find that this is apparently confined to the mainland.

The following seem to be fairly typical—

(*a*) Siberia Jenisei Nikandrovsky ostrov 70° 20′ N. lat. J. Sahlberg.

(*b*) Lapp. Ross. Ponoi. J. Sahlberg and A. J. Malmberg, and also collected at the same locality by R. Enwald and C. A. Knabe.

(*c*) Eskimoland. Seemann.

Ruprecht in his *Symbolæ plantarum Rossicarum*, p. 18, records this from 'Ad promeno. Barmin.'

The plant from Nowaya Zemblya is not nearly so elongate, but the leaves are often 3-lobed—the stem is not so fistular.

(*d*) Nowaya Zemblya Nordenskiöld Expedition. Cap. Gusinnoi.

(*e*) Nowaya Zemblya. N. Kriwoschya.

There is a curious slender form from Spitzbergen in the Kew Herbarium, with always 3-lobed leaves.

(*f*) Joss Spitsburgensis. Th. M. Fries.

Ruprecht also describes a var. γ? *minimus* from Kolguev with a slender stem and deeply tripartite leaves. I have not seen specimens of this, but the plant from Spitzbergen would almost fit the description.

Trollius europæus L. Globe Flower.

This plant, which is extremely local on Kolguev, for I only met with it once, and then in seed—on August 27, by the upper waters of the Pugrinoy—grows even there to the height of 1 foot.

Caltha palustris L. Marsh Marigold.

The Kolguev form of the marsh marigold is very distinct as a variety from our own. It never forms a stout, dense plant as with us; but grows in a trailing decumbent way, and I *believe* roots at the joints, though of this I am not certain; also a copper-coloured underside to the petals is the rule and not the exception. It formed large beds of yellow in marshy places in August.

Delphinium elatum L. Larkspur.

That Ruprecht should not have found this lovely larkspur on Kolguev is not surprising, for it is extremely local, and flowers late. I saw it for the first and only time on August 26 at the foot of Honorohur Mount; and on the banks of the head waters of the Pugrinoy, and also by the Kriva Lake. Few of the plants were then in seed. The flowers were, individually, as large as those in our gardens, though the flower heads were not nearly so long, and the average height of the plant was some ten inches.

CRUCIFERÆ

Cardamine pratensis L. Cuckoo Flower.

This plant is distributed all over Kolguev where conditions are suitable, *e.g.* along the edges of streams. But it is smaller than our cuckoo flower, and I never saw it lilac, but always white.

Arabis alpina L. Alpine Rock-cress.

On Kolguev this is very local. I found it only on the barest north-facing clay banks in August and September. Such places had been covered with snow till late, and with the melting snow the clay had liquefied and slid down. Here *A. alpina* grew. Connected possibly with the extreme shortness of time thus left for fructification is the fact that the seed-pods were far developed (sometimes to a length of $1\frac{1}{4}$ inches), in all cases before the petals showed any signs of falling.

Draba rupestris R. Br.

Common on all the sand-hills.

D. alpina L.

This pretty little draba was quite common on the Pesanka plateau; elsewhere it was little represented, though I picked some plants on the Pugrinoy.

Cochlearia danica L. Danish Scurvy-grass.

Abundant everywhere.

VIOLACEÆ

Viola biflora L. Twin-flowered Violet.

I did not find this flower in Kolguev till August 2nd. It was common on all high dry banks covered with light fine grasses, *e.g. Poa arctica.*

CARYOPHYLLACEÆ

Dianthus superbus L. Splendid Pink.

'A much reduced form of this plant.'

In the British Museum is an example of this plant which exceeds 15 inches in height. On Kolguev I found none over 8 inches, and this in long grass. Its flower is loose and straggling, and at the first glance from the marked ciliate petals, looks more like a 'ragged robin' (*Lychnis Flos-cuculi*) than a pink. In Kolguev I only found it in the Gobista mountains on August 26.

Wahlbergella angustiflora Rupr.

Ruprecht separated this Kolguev form from the *Lychnis apetela* of Vahl—of which the type from Greenland is figured in *Flora Danica*, vol. 13—2173—on the ground that is was more nearly allied to Fries' *Wahlbergella* of Finmarck. The specific name *angustiflora* points one difference, and the calyx also is more tubular, less bladder-like than the Greenland form. I found it first on July 12 at Scharok, where it was common on the sand ridge. A plant brought from Novaya Zemblya by Admiral Markham in 1789 is intermediate between the two forms, while in specimens brought from Spitzbergen by Col. H. W. Feilden (1894), the calyx is at least as bladder-like as in the Greenland form.

Stellaria Edwardsii R. Br.

Hooker and Hartmann unite this to *S. longipes. Goldie.*

This little starwort flourished on the high ground wherever there was enough grass to support it.

S. humifusa Rottb.

It commonly grew at Scharok in July, in damp hollows where water soaked down from the clay level.

Cerastium alpinum L.

This was one of the latest plants in flower on Kolguev. I found it always on dry sandy places, never in wet situations, where it is said usually to flourish.[1] Perhaps this applies to the glabrous form: on Kolguev we had only the woolly variety.

Arenaria peploides L. Sea Purslane.

It is not a little singular that this, the sea purslane of our coasts, should grow at one point only, as far as I am aware, on Kolguev. This explains why Ruprecht missed it. That point is the southern outer koski or sand-banks off Scharok. Here there were many lumps of it in full flower (not in seed) on July 29. It covered heaps of seaweed and hydrozoa on which were nests of glaucous gulls.

GERANIACEÆ

Geranium sylvaticum L. Wood Geranium.

This geranium, which we left in flower at Tromsö on June 8, was so much later on Kolguev, where it is very local, that when I found it on the banks of the Gobista on August 27 scarcely a flower had dropped. The tallest plant was perhaps a foot high.

PAPILIONACEÆ

Astragalus alpinus L. Alpine Milk-vetch.

Common on the higher ground.

Oxytropis sordida Willd.

This, with buttercup (*R. acris*), forget-me-not (*Myosotis arvensis*), and *Nardosmia frigida* was the earliest plant in flower on Kolguev. It is perhaps the most abundant of all plants on high dry parts. It runs through many varieties of colour. My note of June 17 runs:—'The bank above, facing the south, was a perfect flower garden. A lovely little dwarf *Myosotis* was in full bloom; a ranunculus very bright; and a very fine papilionaceous plant like a hairy sainfoin in three varieties, pink, white, and purple. But none of these were higher than the very short grass.'

[1] Bentham and Hooker.

ROSACEÆ

Rubus arcticus L. Arctic Raspberry.

I first saw this lovely little pink raspberry on July 20, in the Pesanka peat levels, and by August 18, when we were high up in the Pesanka plateau, I met with a second form, and thereafter found this very abundant and constant in places of the same kind.

The difference between the two forms is as follows:—

1. The type form. Petals 5. The leaflets oval, often nearly round, bluntly toothed, deeply ribbed and veined. Stem tough and wiry. 1-3 inches. Grows in Kolguev always on peat, in company with *R. Chamæmorus*.

2. Petals in a double whorl. Larger, and curled at top. Leaflets sharply toothed, inclined to heart-shape; but obtuse to flatness at the base; with apex sharply pointed: not deeply ribbed or veined. Stem weak and slender. 3-6 inches. Grows always among light grasses on clay of river banks, where it forms large crimson patches. Both were equally well defined, and grew in these distinct situations.

Rubus Chamæmorus L. Cloudberry.

Professor Saweljew says in his paper on Kolguev: 'The Maroschka (*R. Chamæmorus*), the delightful product of the tundras of Kanin and elsewhere, yields no fruit here.' His companion, Dr. Ruprecht, more cautiously says 'rarissime.' They must have been misinformed by the natives. As a matter of fact the cloudberry does ripen everywhere on the island, but much later than on the mainland; though I never thought the berries tasted so well as those I ate in Norway.

The flower, which was out in the second week in June, was all over by the middle of August. On August 4 it had a small green fruit; on August 13 they were big and red, but hard and sour; and by the 25th we were gathering them by the bushel fully ripe.

Potentilla verna L.

This plant is generally distributed. I saw none with spotted flowers.

Sibbaldia procumbens L.

A common plant at Scharok, forming dense masses on the sides of our camp water-course.

Geum rivale L. Water Avens.

This pretty plant, which I have seen dwarfed in a Hampshire water-meadow, and luxurious round a lake in the Solovetskii Isles, was fairly abundant in the river-bottoms of Kolguev, where it grew to a plant of some six inches high. It did not flower well till August.

Dryas octopetala L.

The white dryas of the Derbyshire hills was common on the Gobista Mountains and the Pugrinoy, flowering in the beginning of August.

Alchemilla vulgaris L. Lady's mantle.

This beautiful plant was one of the exceptions to the dwarfed character of Kolguev plants. It grew at least as finely there as in England. It often formed a brilliant green flooring to the damp passes by which the sleigh tracks led from the gullies, where it grew so fast that it kept pace with the melting of the snow. It did not flower till the middle of July.

SAXIFRAGACEÆ

Saxifraga cæspitosa L.

Generally distributed.

S. cernua L.

This is very widely distributed on the east side of Kolguev. I brought back specimens nine inches high, while its average height was six inches. This plant has little buds or bulbs in the axils of the leaves on the upper part of the stem. I shot a young grouse on August 13 whose crop was full of these.

Saxifraga rivularis L.

This was perhaps the latest plant to flower with us. It was September before I picked the first flower by our water-hole at Scharok.

S. Hirculus L.

The marsh saxifrage was abundant on all the damp places of Kolguev, which were golden with it. Until the flower actually opens this plant carries the head and upper part of the stalk drooping, but as the flower opens the head becomes erect. When the petals fall the sepals turn downwards, enclosing the stalk intimately, and the double-beaked seed-capsule then in many instances changes in colour from a vivid apple-green to a vivid crimson.

S. stellaris L. Var. *comosa* Poir.

This little saxifrage was very local. I found it on the Baroshika banks only, I believe.

S. nivalis L.

Abundant near the sea.

Parnassia palustris L. Grass of Parnassus.

I was surprised to find grass of Parnassus extremely local in Kolguev. I gathered it in the swamps of Solnoi Lake on August 12, and nowhere else.

Chrysosplenium alternifolium L.

Grew all along the stream-sides—everywhere.

CRASSULACEÆ

Sedum Rhodiola DC. Rose-root.

Very abundant at Scharok, chiefly on the dry mud of the estuary sides.

HALORAGEÆ

Hippuris vulgaris L.

Abundant in the lakes of the Kriva district; in fresh-water creeks of Bewick and Solnoida Lakes, where it grows in water 2 feet 6 inches deep.

UMBELLIFERÆ

Archangelica officinalis Hoffm.

Having no works on botany with me I had entered this in my notes as common angelica (*A. sylvestris*), but Mr. Carruthers points out its differences. It is very local on Kolguev. I saw 'the remains of a large dead umbellifer' by the Kriva on June 16; I saw one in the mountains just pushing through the soil on June 28. And I found the plant (since identified) in flower by the Kriva Lake on August 27.

'The garden Angelica (*A. Archangelica*) a native of Northern and Eastern Europe, *long cultivated for confectionary*, is not wild in Britain.' —(Bentham and Hooker's *Flora*.)

MONOPETALÆ

VALERIANEÆ

Valeriana capitata Willd. Small form.

Circumpolar.

This small valerian was one of the most widely distributed of Kolguev plants. It flourished by all the water-courses. It shared with *Ranunculus Pallasii* the distinction of being one of the only two Kolguev plants with a strongly marked scent.

COMPOSITÆ

Erigeron alpinum L. Var. *uniflorum*.

This charming little plant, which with us is confined to open high mountains of Scotland, grew very locally and always on high ground. It was most abundant in the Gobista Hills, when I visited these on September 9, and at that time scarcely a flower had gone to seed.

Gnaphalium supinum L. Dwarf Cudweed.

Almost everywhere.

Antennaria alpina R. Br. Alpine Cat's-ear.

Fairly common on all the high peat-ground.

A. carpathica R. Br.

This plant is exceedingly local in Kolguev. I found it nowhere but on Mounts Sieycherhur and Sowandeyi. The genus has been separated from *Gnaphalium* as diœcious; the flowers of some individuals are composed of filiform fertile florets without any tubular males, in other individuals of only tubular male florets. It might easily be overlooked, as its white cottony-looking head has a general resemblance to the pappus of *Erigeron*.

Matricaria inodora L.

This was not a common plant on Kolguev. The first specimen we saw was growing in the turf on the roof of Uano's hut at Scharok, and I later on found a few plants on the east coast.

Pyrethrum bipinnatum Willd.

This was immensely abundant on all the dry sandy ground.

Artemisia Tilesii Ledeb.

We found this in bud by the Pesanka, July 4, but it was not in flower till a fortnight later. It was fairly common on the hummocky peat.

Achillea millefolium L. Yarrow.

This tiny yarrow—2-4 inches—was very local. I found it only on the side of some inlets by the sea on the eastern side. The plant differs from the yarrow of our hayfields by being dwarfed and covered with down.

Nardosmia frigida Hook.

This was not only one of the earliest plants to flower, but probably the most widely distributed plant on Kolguev. It resembled the Coltsfoot (*Tussilago Farfara*) in that the radical leaves appeared long after those of the flower stem. By the time the flower was in seed the base of the stalk was surrounded by a mass of stout leaves. The nests of the shore-lark were lined with the pappus of this plant.

Leontodon Taraxacum L. Dandelion.

The dandelion flowered very late on Kolguev, where it was generally distributed. In very many cases it never seeded. It was one of the few plants still in flower when we left the island under snow on September 19.

Hieracium alpinum L. Alpine Hawkweed.

This fine hawkweed was exceedingly local on Kolguev. I found it only on one gravelly ridge to the east of the Gobista Mountains.

Senecio palustris DC.

In Britain this interesting plant has become so exceedingly rare that at the present time it is said to be found only in a single fen. As I had never seen it before I was delighted when I found it growing to the height of some 1 foot 6 inches in a fringe round Swan Lake. It requires *mud*, and therefore is absent altogether from Lake Kriva, nor did I see it by Soldonoida.

Saussurea alpina DC.

Common on hilly slopes in August.

CAMPANULACEÆ

Campanula rotundifolia L. Harebell.

I found but a single plant of the harebell, viz., on August 27, in the Gobista Hills.

VACCINIACEÆ

Vaccinium uliginosum L. Bog Bilberry.

Common on the peat-bogs.

It has a bluish-black berry. I did not know this bilberry when I went to Kolguev, and used to puzzle much over its stout habit of growth and woody stems. But we ate the berries, and they were excellent.

V. Vitis-Idæa L. Cow-berry.

This was widely distributed over sandy places. It was common by our Scharok camp. It may be distinguished from the preceding by its red berry.

ERICACEÆ

Arctostaphylos alpina Spr. Alpine Bear-berry.

This bear-berry was very conspicuous on all sandy places in August and September, where it formed bright crimson-scarlet patches as its leaves withered.

Pyrola minor L. Common Wintergreen.

I saw this charming little flower only once on Kolguev. Three plants of it were growing on a Pugrinoy gully when I passed on August 28, the flowers pink, and the leaves edged with crimson. This specimen resembles *P. minor* in that the style scarcely protrudes beyond the corolla, and that the petals are concave and close over the stamens. On the other hand, in the marked veining of the leaves, it inclines to *P. rotundifolia*.

GENTIANACEÆ

Gentiana verna L.

This is so rare on Kolguev that I saw it only in one place, viz., in the Gobista Mountains on August 27. But at that date the flowering was over.

PLUMBAGINEÆ

Armeria maritima Willd. Common Thrift.

The common thrift was very abundant, both inland on sandy hills and by the sea. The scaly calyx of this plant was in all cases of a bright crimson-red; so that in the middle of July the plant appeared to be in the flower before it was really so. When the flower did appear it was exceeding small and of a dirty pink; I never saw one which reached the globular shape of our English thrift. Always the calyx eclipsed it.

PRIMULACEÆ

Primula stricta Hornem.

This little primrose was fairly common on the Pesanka plateau; but was little represented elsewhere.

POLEMONIACEÆ

Polemonium humile Willd.

This, a dwarf, hairy form of 'Jacob's Ladder,' 3-4 inches in height, was common on the dry sandy ridges by the sea and inland. I never found it on the peat.

P. caeruleum L. Var.

This, taller and slenderer than the former, and glabrous, grew on the peat, especially on the peat-mounds. It came out about the first week in August, rather later than the last.

BORAGINEÆ

Myosotis alpestris Schmidt. Mountain Forget-me-not.

I found this in the second week in June on the high hills we crossed near the head waters of the Pesanka. I did not again recognise it.

M. arvensis Hoffm. Field Forget-me-not.

This, speaking generally, was 'everywhere.' It was among the earliest flowers out.

Eritrichum villosum Bunge.

A common plant. Ruprecht separates it as *latifolium*, but, as it seems, on very slender grounds.

SCROPHULARINEÆ

Veronica longifolia L.

It is impossible to describe the beauty of this plant as I first saw it on a bank of the Pesanka ablaze with sunlight on August 18. Later on I found it by the Kriva Lake. But in these two places only. The flowers are more loosely distributed than in our garden form of the same plant, and therefore I think more elegant.

V. alpina L.

This came out with us on August 8 in the Pesanka hills, growing among the grass.

Bartsia alpina L.

This mint-like plant was very local. The floral leaves were tinged with purple.

Castillejia septentrionalis L.

A peculiar-looking flower. It was growing in dense tufts a foot high on the Pesanka on July 4, and a fortnight later it was in flower by the sides of a little stream high up on Pesanka plateau.

Pedicularis lapponica L.

This, which may be recognised by its yellow flower, was abundant.

P. hirsuta L.

This, which has a pink flower, was fairly common in damp places.

APETALÆ

POLYGONEÆ

Rumex domesticus Hartm. Dock.

This dock was pretty generally distributed; but in single plants, not in clumps or masses.

R. Acetosa L. Sorrel.

The common sorrel was abundant, growing finest by the sides of the streams. But even here it was of a much reduced character. We found this, until the cloudberry ripened, a very useful form of vegetable food, as we had no other, and it was often difficult to find a leaf more than an inch long. It never seemed to me to be nearly so acid as in England.

Oxyria reniformis Hook.

This unpretending little plant was exceedingly common on the dry mud-banks at the sides of the creeks. By the end of August the flower had given place to seed; and the plant was then more noticeable and very beautiful, as each seed was surrounded by a broad thin scale, varying in colour from yellow to crimson.

Polygonum viviparum L.

This in Kolguev was a very insignificant little plant seldom more than three inches high. Only the flowers at the extreme tip of the flower-head were developed. The place of the lower ones was taken by little brownish bulbs.

BETULACEÆ

Betula nana L. Dwarf Birch.

This creeping birch was widely spread on all the higher stony or sandy ground. Wherever a wide patch appeared in the distance coloured in brightest emerald, it was found on a nearer view to be covered with this plant.

SALICINEÆ

Salix lanata L. Woolly Willow.

This willow, which grows on the mainland to the height of six or seven feet, is nowhere on Kolguev higher than a man's knee, and reaches this height only in one or two places.

S. Lapponum L. Downy Willow.

More local than the last.

S. Myrsinites L. Whortle Willow.

S. reticulata L. Netted Willow.

S. herbacea L. Dwarf Willow.

This little willow was in the catkin at one spot or another from the time we first landed, June 18, to the end of August.

S. polaris Wahlenb. Polar Willow.

EMPETRACEÆ

Empetrum nigrum L. Common Crowberry.

This on Kolguev was confined to high dry hills. It was in berry at end of August.

MONOCOTYLEDONES

JUNCACEÆ

Luzula hyperborea R. Br. Northern Woodrush.

L. Wahlenbergii Rupr.

CYPERACEÆ

Carex rigida Good.

C. stricta Good.

Eriophorum vaginatum L.

Generally distributed; making sheets of white in the damp places. I picked the first bit fully out on July 24.

GRAMINEÆ

Alopecurus alpinus Sm. Alpine Foxtail.

Aira flexuosa L.

This grass was exceedingly late in coming into flower. 'Flowering heads only now developing' is my remark on this species on July 27. It is this grass which is used by the Samoyeds under the name of Stielka as a padding for their skin boots against the cold.

Trisetum subspicatum P. de Beauv.

Poa flexuosa Wahl.

Poa flexuosa. Var. *prolifera arctica.*

P. arctica R. Br.

Festuca ovina L. Var.

Festuca ovina L.

CRYPTOGAMIA

LYCOPODIACEÆ

Lycopodium Selago L. Fir Club-moss.

This club-moss was exceedingly local. I found it only on the sides of two gullies near the Pesanka about four miles from the sea.

EQUISETACEÆ

Equisetum palustre L.

This plant grows on Kolguev only on dry sandy banks near the sea. It formed the favourite food of a young white-fronted goose.

MOSSES

Sphagnum acutifolium Ehrh.

S. fimbriatum Wils.

Cynodontium Wahlenbergü Hartm.

Dicranum elongatum Schw.

D. scoparium Hedw.

Rhacomitrium canescens Brid.

Polytrichum commune Linn.

P. strictum Banks.

Aulacomnium palustre Schw.

Hypnum felicinum Linn.

H. schreberi Willd.

H. stramineum Dicks.

H. uncinatum Hedw.

HEPATICÆ

Jungermannia ventricosa Dicks.

LICHENS

Stereocaulon paschale Nyl.

Cladonia cornucopioides Fr.

C. gracilis Hoffm.

C. pyxidata Fr.

Cladina uncialis Nyl.

C. ranegiferina Nyl.

Alectoria jubata Ach.

Dactylina arctica Nyl.

Cetraria islandica Ach.

Platysma nivale Nyl.

Solorina crocea Ach.

Nephroma arcticum Fr.

Peltigera horizontalis Hoffm.

NOTE ON THE BIRDS RECORDED

Of the forty-seven birds in the list, I obtained specimens of all but the following :—

Merlin, Bernacle Goose, Wigeon, Tufted Duck, Scaup, Goosander, Common Scoter, Golden Plover, Whimbrel, Bar-tailed Godwit, Arctic Tern, Pomatorhine Skua, and White-billed Northern Diver.

Of these the Pomatorhine Skua, Goosander, Tufted Duck, and Arctic Tern were probably casuals, the Merlin, Common Scoter, and Golden Plover migrants. The Scaup and Wigeon would very likely be breeders, the Whimbrel possibly so, and of the Bernacle Geese I have said in its place all I know. I have reason to believe that the White-billed Northern Diver will be found to be more abundant in Scandinavia than has been supposed, and I think it probable that future voyagers may find it nesting on Lake Promoinœ in the south of Kolguev. The Bar-tailed Godwit I have recorded as uncertain. It was the only bird seen under conditions which left the species doubtful.

Of the thirty-three birds which remain I can record eggs or nestlings of all but the following :—

Wheatear, White Wagtail, Meadow Pipit, Snowy Owl, Brent Goose, Grey Phalarope, Temminck's Stint, Sanderling, Purple Sandpiper, Siberian Herring Gull, and Black-throated Diver.

The Wheatear, Meadow Pipit, Temminck's Stint, Sanderling, and Purple Sandpiper were, I believe, not nesters on Kolguev. The White Wagtail may possibly nest there in small numbers; I cannot say for certain. The Snowy Owl was certainly not nesting there last year. I have explained why I saw no eggs or young in the down of the Brent Goose. Although Professor Palmen mentions the Brent with a '?' in his careful list of circumpolar birds (which deserves the attention of every ornithologist), yet it must be borne in mind that since, as he himself remarks, there was little, if any, ornithological information to be had concerning Kolguev, he naturally included that island in his column containing the records of birds from Kanin and the Lower Petchora.

In the following list (R.)=Russian, (S.)=Samoyed. I have ventured to save myself the responsibility of spelling Russian names by taking these as they are given in Mr. Dresser's work on the *Birds of Europe.*

BIRDS OF KOLGUEV

TURDIDÆ

Saxicola œnanthe (LINN.). Wheatear.

Poputchik Podorojnik (R.).

I saw but a single Wheatear. An immature bird. It appeared at Scharok on August 20, during a time of severe gale and thunder from the NW. From that day till September 6, it sat about on the little banks by the shore.

MOTACILLIDÆ

Motacilla alba LINN. White Wagtail.

Bielo Tresogushka (R.).

There was a single pair by the Gusina River on June 24. On August 21 there were five birds, two old and three young, on the shore at Scharok. On August 26 I saw a family of six birds on a little pond in the Gobista country. Later on we had a small influx of young and old birds at Scharok. They apparently arrived on three successive days, and seemed to be working down the coast.

Anthus cervinus (PALLAS). Red-throated Pipit.

I never saw a single individual of this species on the eastern side of the island, and on the western side it was far from common. A nest taken on June 22 was composed entirely of grasses, and contained six eggs.

Anthus pratensis (LINN.). Meadow Pipit.

Stschewritza-lugowaya (R.).

I have no evidence of the nesting of this bird on Kolguev. I first saw the meadow pipit on August 15, when, after a clear time and full moon, we had a visitation of a flock.

FRINGILLIDÆ

EMBERIZINÆ

Plectrophanax nivalis (LINN.). Snow Bunting.

Podoroschnik (R.). *Oo-noint-sa* (S.).

The snow-bunting was established in many suitable spots, chiefly on the coast of the upper western and eastern sides of the island. I took six eggs from a nest on the Gusina on June 21. This nest, like others which we found, was composed of dead grasses and lined with white feathers of the willow-grouse. It was set deep down in a split in the peat of the clay bank which sheltered our tent. We found at Scharok that a pair of these birds had been nesting last year under the wooden covering of a Samoyed grave. During the last week in August a cock bird was constantly sitting on the turf roof of a hut at Scharok and singing splendidly, with a song which in some passages recalled that of the chaffinch.

On the little island of Hornö, by Vardö, at least three nests of these birds had flown by June 11. On the eastern side, I saw a cock bird with food in its bill, and on the western side two lots of young were flying about the rocks. On Kolguev, to the contrary, we saw a pair building under the overhanging snow of a stream bank on June 27, and young birds at Scharok first flew on July 11; and on July 24 I shot a male bird who was changing plumage.

Calcarius lapponicus (LINN.). Lapland Bunting.

Sizìo (S.).

No bird, except perhaps the shore-lark, was more abundant than this on Kolguev.

It was there distinctly a bird of the peat, in the hummocks of which it nested. On June 24, we first found a nest with eggs—four in number. By August 3rd, they were in small lots of a few families together, by the end of that month in large flocks, and by the second week in August scarcely one was to be seen in the neighbourhood of Scharok.

Otocorys alpestris LINN. Shore Lark.

Javronok-snejng (R.). *Sizìo* (S.).

This bird was exceedingly abundant and universally distributed. We found a nest ready for eggs on June 16, and the first one containing eggs (four in number) on June 27, made of grass and lined with pappus of *nardosmia frigida*.

STRIGIDÆ

Nyctea scandiaca (LINN.). Snowy Owl.

Sowa bielaya (R.). *Hei-nîb-chur* (S.).

The snowy owl, there seems good reason to believe, does not nest on Kolguev. The Samoyeds all agreed that they had never known an instance. The snowy owl is always sporadic as a breeding species, but I have no explanation to offer of this consistent non-occurrence as such, which, but for Colonel Feilden's observations on the habits of this bird in Spitzbergen,[1] I should have attributed to the absence of the lemming. I saw our first snowy owl on June 17, the day after our final landing. It was a dark bird, probably a hen. As it skimmed over a little mere it stooped to pick something—no doubt a fish—out of the water. I was watching it though my glass, and saw it drop its feet and secure the quarry. It then sat on the bank looking over its shoulder at us with the prey under its foot. A red-throated diver swam up and down close in front of the bird with its bill wide open, evidently chiding it, though no sound reached us because of the distance. My notes of the occurrence of the snowy owl are too many to give. Its numbers increased towards the middle of July, and thereafter it was daily in evidence. At first, all

[1] *The Zoologist*, March 1895.

we saw were dark examples; the old males came later. The last record I have is on September 13, after which we left. Every isolated distinct mound (those convex circular mounds I have elsewhere described) was a well-used resting place of this species. Some of the castings contained vertebræ of small fish, but the majority were composed of birds' bones. I only once detected the bones of a small mammal. This goes to show that the owl does its sea passage on an empty stomach. I have mentioned in this book a case where a snowy owl tried to take a white-fronted gosling. The Russian and the Samoyed name for this bird both mean 'snuff' or 'sneeze,' and have reference, no doubt, to the bird's hoarse-breathed noise when disturbed.

FALCONIDÆ

Falco peregrinus TUNSTALL. Peregrine Falcon.

Sapsan (R.). *Har-na-wê* (S.).

Until the autumn migration, the peregrine falcon was rare on Kolguev. We came upon an eyrie on June 17. Here four eggs were laid in a hollow scratched in the ground, on a little level in a bank which sloped to the river Gusina. There was not a sign of a casting by the nest. Though seven days had elapsed since its mate was shot the tiercel was still circling over the spot on June 24, apparently unmated. On August 26 I saw an old falcon on Mount Sowandeyi; and she hung persistently screaming about the same spot. During the latter end of August and the beginning of September we had several immature birds on passage; singly, or two or three together.

Falco æsalon TUNSTALL. Merlin.

Derbnic socol (R.). *Irrah* (?) (S.).

The only merlins we saw were on August 16 and August 17. On the former date a female flew two or three times round the tent, and as she passed within two yards of the opening she turned her head and looked in. On the latter date we saw three together flying low down across the tundra.

ANATIDÆ

Anser albifrons (SCOPOLI). White-fronted Goose.

Kazarka (R.). *Seer-grün* (S.).

Among the many white-fronted geese brought in by the natives, I never saw a single individual which I could refer to the smaller race **A. Erythropus** L. There was certainly some variation in size, but I have no doubt that, like the specimens which I brought back, they all belonged to the larger form. The white-fronted on Kolguev was far more wary than the bean goose. A sitting bird would not allow you to get within gunshot before she left her eggs. I may at once say that I never myself identified this bird at the nest, though I have no doubt whatever, in my own mind, that one of the nests we came across during our walk across the island belonged to this species. The four eggs, slightly smaller and rounder than those of the bean, and also of a clearer creamy white, exactly agreed with some brought to me with the old birds by Mekolka the Samoyed. I have reason to believe that, while the bean goose is pretty generally distributed over the whole island, the white-fronted affects rather the north-eastern district.

On Kolguev this species moulted rather later than the bean. Of those we took at the goosing on July 18 none had lost their primaries, and I saw many on the wing. I have described elsewhere in this book the habits of the young bird in the down, which we kept for a few days alive. They did not bring their young down to the sea as early as the bean, but kept them to the lakes.

A. segetum (J. F. GMELIN.) Bean Goose.

Gummenik Nemock (R.).

This is *the* grey goose of Kolguev. In numbers it exceeds the white-fronted goose by at least three to one. We saw some individuals in which both the yellow of the bill and of the feet very strongly inclined to pink; so much so that I could understand an observer who was not acquainted with the marked difference between the character of the head in this and the pink-foot, mistaking these for individuals of the latter species. A few of these birds had moulted their wing-feathers on

July 18. At this date the majority of the young I saw were apparently not more than ten days or a fortnight old, but were very strong runners. The bean geese did not nest on the peat lands, but they were otherwise equally distributed on the grasses of the low flats and on the high hills. We found no nests with more than five eggs in them, and in two cases the completed number was three. On August 31 we saw the first flocks of geese flying south.

Bernicla leucopsis (BECHSTEIN). Bernacle Goose.
Laboo (S.).

We saw on Kolguev but seven bernacle geese in all. Of these, five were together on the banks of the Gusina on June 23. And two in the country near it two days later. The Samoyeds who gave me the name above recognised the birds at once from the picture, and said there were very few on Kolguev, but that they did nest on the Gusina, and there only.

B. brenta (PALLAS). Brent Goose.
Wurràh (S.).

I never saw the nest or egg of a Kolguev brent. No inducement which I was able to offer to the Samoyeds could extract from them any information as to the breeding places of the birds, until the Russians arrived in August, and with their help I got from Uano a reluctant admission that they nested on the southern and north-eastern ends of the island. They hold the bird in almost superstitious regard because of its extreme importance to them as winter food. I believe that they themselves never approach the breeding grounds during nesting-time. For this they gave as their reason that a dog or a gun would 'make them all be spoilt.' The Samoyed dogs are encouraged as bird-hunters, and a Samoyed cannot understand that it is possible to go out without such companions.

The only reason I have, then, for assuming that the brent goose nests on Kolguev, is the word of the natives and the appearance of vast numbers of old and young off the sand-banks in July. The taking of these is fully described under July 18. About two-thirds only of those were moulted out, but those full-winged did not fly. Of the adults, two forms were clearly distinguishable. In the majority, the lower breast and belly were slaty, but a large proportion had these parts light as mother-of-pearl: and there were some old birds in which the light fringe of the slate-coloured breast feathers was so wide that the bird could not easily be referred to either category. In one bird, a male, in my possessson, the neck and the tail feathers are light brown; apart from this, it may almost be called a white bird.

Cygnus bewicki YARRELL. Bewick's Swan.

Lebed. Chipouna (R.). *Horèh* (S.).

I judge it best to give from my diary the first references to these swans.

June 16, just below the Kriva river. 'As we were walking down a pair of swans came flying past quite low down. As they approached, attracted I think, by old Sailor, they turned and began circling round us. We had heard them some way off. The noise they made was not in the least like that of a whooper. It was a kind of "honk" uttered at intervals, and I had thought was the cry of geese. I looked at them with the glass; the sun full on them. They were so close that it was perfectly easy to distinguish the face,'—*i.e.* the shortness and shape of the *orange* (not *yellow*) which looked *red*—'they were Bewicks. I was so much interested in watching them that I never thought of shooting till Hyland woke me up, saying, "Shoot, sir!" I had just time to change one of my dust cartridges for s.s.g., and fired at the second swan as they were getting rather out of range. The bird, struck in the side, fell, or rather flew down slantingly—the other bird flying up and down above it—and pitched about 200 yards away. On Hyland going up it rose and flew on, accompanied by its mate, for perhaps another 300 yards, very low over the tundra. I called Hyland back, saying we should find the bird on the way home. And sure enough, as I passed on our hurried return, there was the swan (within a hundred yards), but we could not then wait.[1] One bird flew off; the other remained in a sitting position. I was terribly sorry about

[1] For reasons explained under this date.

this. Just about fifty yards south of Bewick Lakes, Hyland shot at and crippled in a pond a male long-tailed duck which we had been stalking. We had to leave it after all, for it dived away among the ice-floes among which I feared old Sailor, who had plunged in before we could stop him, would lose his life. As he fired, two swans, one of which I had just got my glass to bear, rose and flew off. I just noticed that the second rose from the top of a big mound. I had not dared to talk, and could not signal to Hyland. We were on opposite sides of the pond. I knew them for Bewicks by their character and small size. I had not time to identify them by the head. They made no call as they went. The mound was a nest about 2 ft. 6 in. in height and 4 ft. 6 in. in diameter at the base. [It was perfectly smooth and symmetrical, tapering till the circular top was no more than about two feet across.] This structure was entirely composed of little bunches of green moss, with the exception of a very little lichen, and a chance bit here and there of short light dead grass pulled up with the moss; of course there were no green grasses or reeds as yet, and not a single piece of dead reed had been used. There was a thin lining only to the nest of dead grass *mixed with* a little down. Some, but not nearly all of the moss, had been pulled up near the nest, which was situated on a dry place, rather grassy than mossy, till one looked close. There were three uncovered eggs in the nest, and a broken one lying on the ground beside it. Of these three I took two, hoping that the birds would continue laying. I wish I had covered up the egg, but I hardly liked to do so. Three skuas (*S. crepidatus*) had flown off from near by as we approached. We left, and having gone a very short way, turned, and, looking towards the swan's nest, saw an Arctic skua sitting ominously about on some tussocks near it, and when we looked again it was sitting on the edge of the nest and hammering downwards. So deep was the nest that the bird as it hit on the egg almost went out of sight. Had it not been for my anxiety to return I should have gone back, but it was just at the critical moment of the change in the wind.'

The eggs I have exactly agree in texture and measurement with those of Bewick's swan as given in Yarrell.

I have also described the killing of the young swans with bows and arrows on August 10. The beaks of those birds were of a light slate colour, through which was visible a red vascular network, so that the general effect was that of a pinky-slate.

A Samoyed afterwards showed me the skins of seven adult Bewick swans. We saw nothing of the whooper.

Mareca penelope (LINN.). Wigeon.

Swetsun (R.). *Harlgàr* (S.).

I can only record a single pair of wigeon. These I saw on a small pool near the Sauchika entry on June 17.

Fuligula cristata (LEACH). Tufted Duck.

Tschernett (R.).

On August 11 we had a female tufted duck on a little lake before our Pesanka camp. She always kept to the middle, just out of shot. Twice when I was collecting insects at the edge of this lake she came flying close past me on her way to a second pool. Her identity is beyond all doubt.

F. marila (LINN.). Scaup.

Goloubaïa, Tschernett, Beloglaska (R.). *Harocèrl* (S.).

On June 16 we saw a pair of scaup on one of the Bewick Lakes; and on July 8 I saw another pair on a lake in the lower Pesanka district.

Harelda glacialis (LINN.). Long-tailed Duck.

Vostrohwostka (R.). *A-ðo* (S.).

Out of the numerous gatherings of these ducks we saw but a single drake in full spring plumage. This was on the sea on June 15. But though the drakes had lost their white heads and white scapulars they retained their central tail feathers. These were not moulted out till the first week in June, and on July 8 my note runs: 'One scarcely sees a single male now with a long tail.' Again, July 22: 'Saw several harelda; only one drake with a long tail.'

This duck far outnumbered any other species on Kolguev. The first nest I found was in the interior of the islands on June 27. No water of any kind could be seen from the point where the nest was situated. It was rather high up, and on the peat. 'Moved harelda off her nest. Quite away from any water (as far as we could see) this nest was remarkably deep and neat; all of down, with a very little dead grass and dead birch-leaves. It contained six eggs, which I took. They were slightly incubated.'

Besides the common call so well expressed in the Samoyed name, this bird has a remarkably human cry. It would be hard to overstate the wonderful diving powers of this bird. On July 30, when trying to work some long-tailed ducks across a small lake within reach of Hyland, I fired a shot which made some little ones of this species who were with their mother dive. The old bird did not dive but swam straight away, and, watching through the glass, I could see the young disappear, and then after a bit appear again one by one *close to* the old one, and instantly dive again; and so, diving and reappearing, they kept up with her all across. When we were lying in the Kolokolkova gulf (on the mainland), in spite of the tremendous gale, which was such that the boats could not venture out, there was a party of three ducks within sight of us for some hours. They would dive to meet only an unusually big-topped broken wave: ordinarily they paid no attention to the broken water which swept over them.

Somateria spectabilis (LINN.). King Eider.
Pistrak (R.). *Na-ma-tǜr* (S.).

The king eider is, after the long-tailed duck, the most abundant species of duck on Kolguev, though it only bears to the latter the proportion of, say, about one to twenty. We saw a great many on the sea during those days on which we were sailing up and down in front of the ice. I got a pair by the Kriva on June 16—the male in magnificent plumage. On the eastern coast we saw them also from the outer sand-banks in considerable quantities flying up and down the coast. But though they *crossed* the mud-flats occasionally on their way to points above or below, we never, in any instance, knew a male bird settle in the tidal creeks. With the exception of the bird above referred to I never saw a single male of this species inland. The nests of the king-eider were generally distributed from the crest to within five or six miles inland. The Samoyeds constantly brought in the eggs during the end of June and beginning of July, and also shot the females.

July 19.—'After its nest had been passed within, I should say, a *yard*, a dozen times during these two hours by dogs, men, sleighs, etc., a duck eider suddenly flew straight off her eggs—twenty paces exactly from where I was sketching and the women were stowing the geese—and away to the creek. Yet these eggs were all carefully covered up. They were five in number, and considerably incubated.'

This was a late nest. The date for the appearance generally of the young on the lakes may be put about the 21st of July. At this date every little lake held a duck or two with young ones. They are much more wary than the long-tailed ducks, and very hard to drive.

This is an excellent bird to eat; whereas the long-tailed duck is very poor food.

We saw nothing of the common eider on Kolguev. The Samoyeds did not recognise its picture, though they knew the king eider's at once. I examined carefully very many of the birds brought in, but they all, like those I have brought home, were unmistakably king eider ducks.

Mergus merganser (LINN.). Goosander.
Krahal dlinnonosoy (R.). *Harrl-gart* (S.).

We saw three goosanders near the harbour entrance on July 25, and a party of eight of these birds were off the sand-banks on July 29.

Œdemia nigra (LINN.). Common Scoter.
Chernaya iitka (R.). *Si-è-gle-via.*

On July 26 a big flock of these birds came in with the tide and were diving off the sand-banks, 'the leading bird disappearing first, then the next, and so on in order right out to the tail. Then on rising they would all close up together, and again string out and dive as before.' I saw nothing else of the scoter.

TETRAONIDÆ

Lagopus albus (J. F. GMELIN). Willow Grouse.
Koropatka (R.). *Hòndiii* (S.).

My first notice of these birds, taken from my Journal for June 16, the day we landed by the Kriva, runs—'A grouse was very abundant. The males were most conspicuous objects, and at one point near Bewick Lakes we had twenty of them in view from the same position. The hens as a rule sat [*i.e.* squatted] very close, almost until trodden upon. The males either flew wildly and straight away, or else from point to point within a distance of thirty or forty yards round us. Each time, before settling, they rose with fixed wings a few feet up to a point, like a toying wood-pigeon, and then shivered their wings at the point, and, still shivering, settled.'

On June 17 I found a nest containing twelve eggs near Sauchika entry. On August 6 I obtained a cock bird who was changing plumage, some red feathers appearing on the back and claws just ready to be shed. On July 30 saw eight cocks together, and from then onwards they continued packing.

I did not notice any young till August 3. I came across, on that day, two broods of cheepers who could fly thirty yards or more, and on subsequent days found many. The chicks of a brood were then of very unequal size. These broods were always amongst the willow, but earlier in the summer we had chiefly found the bird on grassy patches among peat. On August 14 found a brood in which the young were nearly as big as the old.

I was surprised to find that these birds do not cross to the mainland in the autumn. The Samoyeds assured me that on Kolguev they formed an important winter food of the foxes.

CHARADRIIDÆ

Eudromias morinellus (LINN.). Dotterel.
Zuek glupöi (R.).

I saw a single bird on June 22 near the Gusina, and no others but those referred to in the following note. July 4, 'Soon after rising the Pesanka hills we halted and pitched on a highish bit of ground overlooking the river and some lakes. I went off at once with Hyland, and we found a dotterel's nest with four eggs. The old bird ran off lame.' Also some 200 yards from the nest I saw a second bird. The following day there was a bad north-westerly gale, with snow, but on July 6, after watching for a long time to see whether a second bird would show itself, I shot the nesting bird, took the four eggs, and, after going some little distance, perhaps 200 yards, I shot a second bird, but could not find its nest, though it was evidently resting. Both these birds proved to be females. These four eggs appear to belong to a set: they present no noticeable differences.

The Samoyed group a great many of the waders together under the name 'Sierrk,' which is evidently a corruption of the above Zuek. It is curious to find, by the name given by Mr. Dresser, that the Russians call this the 'foolish' dotterel—Zuek glupoi.

Ægialitis hiaticula (LINN.). Ringed Plover.

This, the *smaller* form of the ringed-plover, was to be met with in many appropriate places, from the coast to the highest inland hills. On June 24 we found a nest with four eggs.

Charadrius pluvialis (LINN.). Golden Plover.
Rjanka sivka (R.).

On July 29 I saw a solitary golden plover with half-black breast. This is the only occurrence of which I am certain. I fancy two birds which I recorded on the Kriva as golden may have been grey plover.

Squatarola helvetica (LINN.). Grey Plover.
Rjanka Tooles (R.). *Doo-lissia* (S.).

From the day of our first landing till about the end of the first week in September we almost daily saw the grey plover. The first nest I found was on June 26 on a high down about half-way across the island. 'Here I took a grey plover's nest of four eggs and shot the female. The male of this species is exceedingly wary and wild. They very seldom venture within shot. After waiting about for a long time and sending Hyland round also with his gun, I had to give it up. There is no possibility of confusing this with the golden plover, even when on the wing. The call is quite different, and I think the flight more powerful, and that is saying a good deal. The skuas stood no chance with them. They actually seemed to both of us to hit the skuas, wheeling round them and then, making a point high above, they would drop down like a stone, literally knocking the skuas out of time. The nest was a *deep* circular depression, and contained nothing but the eggs and a little lichen.'

In another nest on July 13 the eggs contained fully formed young ones. From August 10 onwards there were immense flocks of these birds constantly wheeling over the mud-flats. These birds behave very differently at different times when nesting. Sometimes the hen-bird feigns lameness, though I never saw the male do this. Often, however, their actions exactly recall those of the stone-curlew, excepting that we never found a male brooding the eggs. The male bird, who always sits on some raised point at a little distance from the hen, gives, long before you come up, an alarm signal to the hen, whereupon she runs off the nest and joins him. The breasts of the males we shot were all equally black, but those of the females varied a great deal.

Strepsilas interpres (LINN.). Turnstone.
Severnaya-Kamnesharka (R.). *Tŭ-ik* (S.).

June 17. 'Turnstones were, especially near the lakes, fairly numerous. I am sure from their behaviour that some of the pairs had nests, though we failed to find them. They were singing a very charming little song which went, as I took it down at the time, 'Chĭwāh chĭwāh chĭwēēkĭ, kĭ kĭ kĭ, kī kī.'

During the nesting season these birds are distributed far inland. On July 12 I found four tiny young ones which 'ran out of the nest in all directions.' The turnstone, like many of the waders, makes several scrapings or false nests before it finally lays.

While the reindeer-salting was going on at Scharok during the first week in September, a pair of these birds were constantly occupied in turning over the contents of the deers' paunches which lay about.

SCOLOPACIDÆ

Phalaropus fulicarius (LINN.). Grey Phalarope.
Plavunchik-plosconosey (R.).

I saw but a single pair of these birds, near the Kriva on June 16.

P. hyperboreus LINN. Red-necked Phalarope.
Plavunchik-cruglonosey (R.).

This was a very familiar bird on Kolguev, and was generally distributed. Every little lake in which there was any vegetation held its one, two, or three pairs. The lobing of the feet was well-marked in even the tiniest young which Hyland found on July 12. The stomachs of some we examined on June 16 were filled with larvæ of the musquito. The following from my diary of July 12 is worth quoting. 'A little stint which I moved to-day kicked her tiny young ones away. The action was so evidently intentional that I spent some half-an-hour in experimenting the point. Six times I let her settle down to brood them, and six times I moved her, and always she did the same. Once she kicked only one away, four times she kicked two, and once three—one a few inches to the side of the spot, and two a long foot on either side behind her. The little ones instantaneously recovered themselves, for sometimes they were lying on their backs, and squatted absolutely still. Then on returning she would brood first one and then another, until she had called them all together again. When Hyland joined me I was on the point of mentioning this to him when he said, talking of a red-necked phalarope's young which he had found, "That old phalarope got off in such a hurry that she knocked two of the young ones right head over heels."'

I had been struck last year when in Norfolk by exactly the same action on the part of a ringed plover. My note made at the time, after describing the case, adds: 'It could not be clumsiness, because these birds are exceedingly careful when moving off their eggs. Gallinaceous birds and ducks, when disturbed and starting suddenly off, are very apt to jerk an egg or two out of the nest—

probably those which have been lying on the feet. But they are birds which fly straight off the nest with considerable flurry, whereas the waders step off quietly.' By the end of July there were great numbers of these birds swimming on the sea in lots that varied from eight to twenty.

Tringa alpina LINN. Dunlin.
Pestrosoboy-pessotchnik (R.). *Yà-i-bud-dy* (S.).

The dunlin was generally distributed. They were in pairs on June 16, but not yet nesting. One bird obtained on the date contained a half-formed egg. 'They were occupied in feeding and in drumming like snipe, but with a more metallic sound (like a pea in a tin whistle) evidently made by the voice.' On July 6 I took four eggs 'hopelessly incubated.' On July 9 I found young ones. Hitherto these birds (which principally affected those districts where the peat was hummocky) had been seen in pairs; now the males (?) began to get together in lots of a dozen and more. And from July 11 till end of August there were immense flocks on the mud-flats of the eastern coast. On August 31 these birds were flying south.

As in the case of the ringed-plover, the dunlin of Kolguev was the smaller form.

T. minuta LEISLER. Little Stint.

The little stint was by far the most abundant wader on the island, and next numerically was the dunlin. On June 16 they were in small parties by the Kriva, chasing one another about, and none appeared to be nesting. On July 27 we found a nest with four eggs. It was a cup in the peat half filled with dead leaves of creeping birch only. In the many nests I examined there were always dead leaves in the bottom of a cup—leaves generally of the creeping birch (*betula nana*) or, according to the surroundings, of *vaccinium*. Seldom had any other material been used. But a nest found on July 9 was 'a deep cup lined with dead birch leaves and a little dead grass.' We never in any instance saw more than one bird at the nest, or with the young. Out of seven birds secured under these circumstances, five were females, two were males. During the nesting season and up to about the fifth week in August, parties of little stints numbering five, ten, or fifteen birds or so, might certainly be seen flying in and about the lakes. When flying in this way, these birds look very like larks and make a twittering noise. No words could adequately convey a good idea of the complicated ruses adopted by this tiny wader near the nest: nor could one express phonetically the various notes it then utters. Thus: 'July 10, The little stint who had the nest made while feigning lameness, etc., a noise exactly like the squeaking of a house-mouse.' 'July 12, Took one nest of little stint with four eggs. One bird only at the nest as usual—a male. It behaved, as H. said, like a dancing doll, jumping up and down on the same spot as if on springs.'

T. temmincki LEISLER. Temminck's Stint.

I saw very few of these birds on Kolguev, and none that seemed to be nesting. I shot a female that was alone on the Scharok mud-flats on July 16. I saw three together by a gully stream on Pesanka plateau on August 7.

T. striata LINN. Purple Sandpiper.
Pesoschnik morskoi (R.).

The first occurrence of this bird was on June 22, when I shot a pair high up by a stream in the Gusina country. On two occasions we saw a single bird, once two, once three, and once five together on the sands of the eastern side of the island. The last of these occurrences was on September 8. They showed none of the solicitude of nesters.

Calidris arenaria (LINN.). Sanderling.
Morskoi-sujok (R.).

The first we saw was a bird I shot off the sands at Scharok on July 11. It proved to be a male and was in summer plumage. On August 27, large parties of the birds were feeding at the edge of the tide on the outer sand-banks: an occasional individual retained some rufous feathers on the back, but in the great majority only black feathers showed against the grey.

When we reached the Kolokolkova gulf on the mainland, great numbers of these birds were running along the edge of the water, and the sands were spotted in scarlet from their droppings. I found that they were feeding greedily on a pteropod (*Clione limacina*) and certain medusæ (*Turris digitalis* and *Sarsia* sp.), and that the brilliant colour which made the sands appear at a little distance as though covered with scarlet geranium-petals was due to the pigment in the generative organs of the hydrozoan.

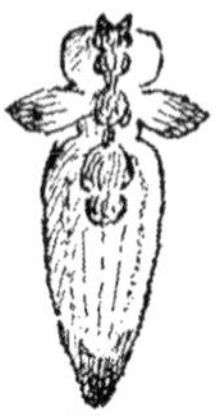

Clione limacina.
The sanderling's food

Machetes pugnax (LINN.). Ruff.
Dratschounn (R.). *Lorr-sōw* (S.).

By the Kriva on June 16 'we met with reeves at intervals, and came upon a ruff's playing-ground, on which twenty ruffs were engaging one another in a most grotesque manner. We could only see two reeves near this ground.' On June 28 I found a nest containing four eggs. On July 26 I came upon a reeve with young about a week old; I saw males in winter plumage on various dates up to the middle of August.

Numenius phœopus (LINN.). Whimbrel.
Kulik (R.). *Arki-sierrk* (S.).

The Samoyeds recognised the whimbrel's picture, and said they called the bird by the name above, which means 'big wader.' Our acquaintance with it on Kolguev was limited to a single pair, which flew overhead by the Gusina on June 19.

Limosa lapponica (LINN). Bar-tailed Godwit (?)

I *believe* I saw a pair on June 16. 'I also saw a pair of large waders, which flew wildly off. I could not identify them positively, but believe them godwits.'

STERNIDÆ

Sterna macrura NAUMANN. Arctic Tern.
Krashka-morskaya.

While we were lying off the west coast of Kolguev in the beginning of June we saw a few of these birds. But I do not believe it nests on the island. We never saw an example after we landed.

LARIDÆ

Larus affinis REINHARDT. Siberian Herring Gull.
Tschaika (R.).

There were always a few of these birds at Scharok. They nest on the low cliff, and in spite of constant robbery by the Samoyeds, manage to get a few young ones off.

L. glaucus O. FABRICIUS. Glaucous Gull.
Morskaia Tschaika (R.). *Har-lèy-o* (S.).

The nests of the glaucous gulls which we visited were situated on the highest ridge of the outer sand-banks to the south of Scharok harbour. They were visible from a very long distance, and proved to be lumps formed of sand mixed with sea-weeds and great quantities of hydrozoa (*sertularia* and others), on which flourished *arenaria peploides*. The sand had in many cases originally collected round drifted timber, and the birds had taken advantage of this to raise upon it a pile some two feet and more in height. As the Samoyeds rob these nests constantly one wonders that any young get off. Hyland was so violently mobbed by these birds, which stooped right down at his head, that he shot two 'in self-defence.'

Pagophila eburnea (PHIPPS). Ivory Gull.

The ivory gull is, according to the Samoyed, a winter visitant to Kolguev. I picked up a wing belonging to this species.

COLYMBIDÆ

Stercorarius pomatorhinus (TEMMINCK). Pomatorhine Skua.
Pomornik scrednic (R.). *Hourioù* (S.).

We saw a few examples of the twist-tailed skua near the sea, but I have no reason to believe they nest on Kolguev.

S. crepidatus (J. T. GMELIN). Arctic Skua.
Pomornik-tschujeadnui.

We saw a great deal of the Arctic Skua. I have spoken of them earlier in this book in connection with Bewick's swan and the grey plover. Although greatly dependent when at sea upon the labours of other gulls, the breeding pairs are as persistent robbers of eggs as rooks in a dry season, and may be constantly seen quartering the tundra for eggs or young. I should be inclined to estimate that of breeding birds on Kolguev there is about one pair to every seven square miles of country. We never found a colony, nor even two pairs together. All those I saw belonged to the light-coloured race. On June 29 we took eggs about half incubated. The nest was among dead water-grass in a bog, and was more than a mere depression for grasses had been walled into a lining. A nest containing one egg (July 7), was a simple depression in dry grass: this egg had a remarkable escape. We were driving along—four sleighs, which meant eighteen reindeer—when I called out to Hyland who was in front to stop; for some 30 yards or so away, a pair of skuas were behaving as though they had a nest. However, we could make nothing of it, and had just taken our seats again to start off, when as I stooped down to disengage the hind leg of one of my deer, lo and behold, there was the nest under my sleigh. The whole train of sleighs had passed over this nest, and yet the single egg was not broken. On August 7. 'We picked up a young skua and brought it back alive. This bird was almost full grown and had well-developed primaries. Its parents showed no anxiety about it. It was beside a lake, and as we approached ran and hid in some grasses. It bit viciously, but made no noise.'

I never in any single instance knew an Arctic skua stoop at a visitor near its nest. On the contrary, an intrusion was met by every wile of allurement. It was the old game of 'hot or cold;' until at last, when you stood close to the nest, both the birds were reduced to a state of helplessness. At such a time they behaved exactly alike. Sitting on their tails, either in the water or on the grass, and beating forwards with their wings, they mewed all the time like cats.

Colymbus adamsi R. R. GRAY. White-billed Great Northern Diver.
Morskaia Gagàra (R.).

One of these birds passed within a few feet of where I was standing on the *Saxon* on June 15, off the western coast, and I saw one flying seaward on June 23. This concludes my record of the species.

Colymbus arcticus (LINN.). Black-throated Diver.
Gagara polosataya (R.). *Pa-du-o-rèh dei-nù-nya* (S.).

By an explicable chance we never found the nest of this bird. They nest late, and when we were at the nest places, the reeds were not sufficiently grown. But we saw the bird on the lakes, and it was almost daily in evidence, fishing in the creeks at ebb-tide. I have before referred to a noise it makes like a dog's yelp, and may add that, when undisturbed on the lake, it makes a noise like the croak of a bullfrog. When flying high overhead, the bird frequently utters a cry so like the crow of a cock grouse that we were often momentarily taken in by it.

C. septentrionalis (LINN.). Red-throated Diver.
Pa-du-o-rèh Pei-o-rèh (S.).

The red-throated diver outnumbered the last species by five to one. Almost every little lakelet held its pair. July 9.—'I took by the lake a red-throated diver's egg. The nest was a shallow depression in a heap of grass and waterweeds in water ten inches deep, with two well-defined tracks running up to it. Down one of these the sitting bird dived, like a running pike, straight out into the lake, so that for many feet I could see the cut of the water as it went.' Aug 11. A bird in the down was brought me by a Samoyed, who declared that it ran out on to the ground when pressed. If it had not been told me by a man whose word I had the strongest reason to trust, I should not have quoted the statement.

A monograph might be written on the ways of this bird and the last: which were an unfailing source of interest and wonder to me. But the following may be a useful note, for the noises of the divers are many and confusing. July 22. After a whole morning by our Scharok lake. 'The notes of the red-throated diver when on the lake are reducible to two forms: (i.) A long-drawn-out *cry* somewhere between a human wail and a cat-call (which we came to talk of as "mewing"); and (ii.) a short of churning bark. The latter is made by one of the birds (I *think* by the female) when making up to the other. It is done by fits. A pair of birds will be swimming along one behind the other, when suddenly the leading bird will face round, lay its head down on the water and churn.' The noise can be heard at a considerable distance.

NOTE ON THE MAMMALIA OF KOLGUEV

THERE is no rodent on the island. I had expected to find *Myodes lemmus* L., the Western Lemming, but it was not there. Equally surprising was the absence of *Lepus variabilis* L., the Arctic Hare. The Samoyeds had never known of an instance of the occurrence of either of these animals.

The theory of the recent origin of Kolguev gains in evidence from this fact. For the extent of water or ice between the mainland and Kolguev is sufficient to prevent the passage of both of these creatures.

We were so little by the sea that my notes on the whales are of small value, and the same applies to the seals. The seals which most commonly occur are apparently *Phoca grœnlandica* Nilss., the Greenland Seal, which the Samoyeds call *Nierpi*, and *Ph. vitulina* L., the Common Seal—*Nhuk* in the Samoyed. Kolguev does not seem to be a very good sealing-ground. The Russian traders took from the Samoyeds very few skins. But these natives obtain sufficient for their own requirements. On Kolguev boots, gloves, and harness (except the saddle and collar) are made of seal-skin. Seal-skin also takes the place of rope; for thongs of considerable length are obtained by cutting a continuous strip round and round the animal's body.

Trichechus rosmarus L. Walrus.
Morse (R.). *Di-ù-li-eh* (S.).

The walrus is only taken by the most adventurous of the natives. Their boats are really not fit for this hunting, nor for going up to the ice. Only two walrus were killed last year.

Ursus maritimus Desm. Polar Bear.
Oskoog (R.). *Hèv-diii* (S.).

A chance polar bear or two appear each winter on the island. In the winter of 1893-94 Uano killed a female and two cubs far inland in the Pesanka hills. The dogs bay the bear and the men shoot it with flint-lock rifles.

Canis lupus L. Wolf.

The wolf very rarely crosses to the island. But in the winter of 1892-93 two came over the ice, as Uano told me, and had killed twenty-five of his reindeer before he shot them.

C. vulpes L. Fox.
Lisitsa (R.). *Tchu-òr-nia* (S.).

The red fox is not very plentiful. During the summer it frequently lies out on the moss or in the willow bushes. It is a lighter-coloured, lankier animal than our own, and has a longer brush and flatter sides. It chiefly preys on the willow-grouse.

C. lagopus L. Arctic Fox.
Psi-itz (R.). *No-hò* (S.).

This is far more abundant than the last species. These foxes live colonially, in warrens in the dry sand. They feed principally on remains of dead seals, fish, crustaceans and molluscs, which they find on the shore. I have reason to believe that they also take live fish from the lakes.

Cervus tarandus L.
Oleyn (R.). *Tii* (S.).

I was never able to determine whether wild reindeer ever cross to Kolguev from the mainland. There seems no reason against their doing so; but the Samoyeds disagreed on the point. Some said they did, others maintained the contrary.

I was assured by Alexander Samarokoff and his cousin that in the days before the reindeer plague there were no less than 25,000 reindeer on Kolguev. Most of these were owned by Russians. At the present time there are on Kolguev 2740 adult deer, and they are owned as follows:

Alexander the Russian and Uano the Samoyed jointly own	400
Uano himself owns	200
Alexis the Russian and On Tipa the Samoyed jointly own	200
Alexis himself owns	500
On Tipa himself owns	800
Yelisei the Samoyed owns	100
Alexander's nephew and Vasili Popoff (a Russian) jointly own	500
The brothers Bulchikoff (Samoyeds) own	40
	2740

The value of a live reindeer is 10 roubles=£1. 'He who has reindeer has everything,' says a Russian proverb. It is true. Without their reindeer these Samoyeds could not exist; with them they cannot want. Almost every part of the animal is of use. The hide makes clothes, bedding, and parts of harness. The bone makes arrow-heads, buttons, and thimbles. The sinew is used for cotton and thread. The horn makes arrow-tops (grooved for the string), powder measures, parts of harness, and buttons for toorrs (toorr-mahl). The flesh is eaten, and the animal and its parts have an exchange value.

Travellers in Siberia and the east of the mainland tundra have assumed that a practice with which they are familiar there is general. I refer to the removal of one of the horns of each draught reindeer. This practice does not obtain on Kolguev, nor is it to be met with, I believe, to the west of the Petchora. Out of the very many hundreds of reindeer which I saw on the mainland not one steer had a horn removed; or only in exceptional cases of 'snow-shovels' of abnormal growth. There is indeed no good reason for the practice. For the deer of a team have no difficulty in so carrying their heads that their horns shall not be entangled. Only very exceptionally does a deer become impatient in this respect. Ordinarily (and it is extremely interesting to watch) they arrange the relative positions of their horns by mutual compromise.

The horns of the bulls are often cut off entirely. The horns of none of the Kolguev deer were clean until the last week in August. And many were still ragged in the middle of September. Many of the gelded animals never lose the velvet, though they regularly throw their horns. The bull whose horns were clean began rutting at once. And this is curious, because they had not nearly recovered their thick coats. In this respect the Kolguev deer are much later than those on the mainland. I saw no bulls with full necks till we reached the mainland (Sept. 27), where we found the majority so.

A bull is not allowed to breed after he is five years old, but a cow will breed up to fifteen, and sometimes even to twenty years.

After the cows have done breeding for the season they are not infrequently used for sleigh-work, especially by the smaller owners.

A beast is best for sleigh work at four years old.

The calves are dropped in May and June, and are entered to sleigh work at ten months old. They are trained in a team of five together; the principle of putting them with an older beast for steadiness does not obtain.

Besides 'the fly,' reindeer have an enemy in the musquito, which, especially during moulting time, punctures the veins of the face.

CLIMATIC CONDITIONS OF KOLGUEV

SAWELJEW, in the interesting paper to which I have already referred, makes the following remarks about the climate of Kolguev:—

'During the sixteen days which we spent at various spots on (or off) the island in July and August, the thermometer never rose above 9° R., and even this but once, at mid-day. Usually it stood at 4° or 5°, and fell at times to 2°, or even to 1° R.; while, before this, on Kanin (*i.e.* on the mainland), a warmth of 10° to 12° prevailed, which, immediately after our departure from Kanin, mounted (there) to 15° R. It is to be remarked that the soil of this island . . . does not thaw in the course of the year more than two feet deep. Further down all remains in a frozen state—a thing which we have not found to occur either on the Kanin peninsula or on the coast of Timan.'

Of course this last remark refers to boggy ground. I was unfortunately unable to test the conclusion.

The Russians have a great dread of being compelled to pass the winter on Kolguev. They told me that only one Russian to their knowledge had ever done so. This man was Alexis' father, who postponed his departure till too late. He described the winter as terribly severe; and the Samoyeds all agreed in saying that it was far worse than on the mainland.

The average shade temperatures for the twenty-four hours during June were, max. 40° F., min. 33° F., the thermometer twice falling to 31° F., and twice rising to 50° F. On June 24 it registered in the sun at mid-day 62° F. The average for July was decidedly higher, though more than once it fell to freezing-point. On August 28 my thermometer was broken during a gale. But up to that date it varied from 42° F. at noon to 76° F. This, the highest reading, was on August 16, and on this day it was 86° F. in the sun. The thermometer never fell below

42° F. during August up to the 28th. After this the weather got rapidly colder, and by September 16 the ground was covered with snow, and the lakes were beginning to freeze.

Kolguev is exceedingly subject to fogs and gales of great duration. The prevailing wind is northerly, either N., N.E., or N.W. Indeed, this is so remarkable that I give the record:—

June 13.	S.E. to N.E.	Light.
„ 14.	N.	Fresh; snow falling on hills.
„ 15.	N.E.	Light; dripping fog in early morning
„ 16.	N.E.	Light; clear.
„ 17.	S.E.	Light; clear.
„ 18.	N.W. to S.W.	Light; dripping fog.
„ 19.	N.W. to N.	Strong; thick fog; rigging ice.
„ 20.	N.E., N.W.	Strong; clear, sunny.
„ 21.	N.W., N.E., E.	Strong; dripping fog.
„ 22.	N.E.	Light; clear, sunny.
„ 23.	N.E. to N.	Light; clear, sunny.
„ 24.	N.E.	Light; fog.
„ 25.	N.E. to E.	Strong; froze at night.
„ 26.	N.E.	Strong.
„ 27.	N.E to E.	Light; froze at night.
„ 28.	N.E. to S.E.	Light; dripping fog.
„ 29.	N.W. to N.	Light; dripping fog.
„ 30.	N.W.	Light; dripping fog.
July 1.	N.W.	Strong; dripping fog and freezing.
„ 2.	N.	Very fresh.
„ 3.	N.	Gale; snow.
„ 4.	N. to N.W.	Light.
„ 5.	N.W.	Gale; snow and rain.
„ 6.	N.W.	Gale.
„ 7.	N.W.	Less wind.
„ 8.	N.W.	Fresh; sunny.
„ 9.	S.W.	Light; sunny.
„ 10.	S.W. to N.W.	Strong; rain and dripping fog
„ 11.	E. to N.W.	Very light.

July	12.	N.W.	Very strong to gale at night.
,,	13.	N.W.	Very stiff gale.
,,	14.	S.W.	Light; sunny.
,,	15.	S.W.	Light; lovely day.
,,	16.	S.W.	Fresh; dripping fog at night.
,,	17.	N.	Fresh; dripping fog all day.
,,	18.	E.	Fresh; rain all day; night, dripping fog.
,,	19.	S.W. to N.W.	Light; clear.
,,	20.	S.E., S.W., N.E.	Light; dense dripping fog.
,,	21.	N.E.	Strong; rain.
,,	22.	E. to N.E.	Gale; dripping fog all day.
,,	23.	N.E.	Strong; clear.
,,	24.	S.W. to N.W.	Blowing hard all day.
,,	25.	N.W.	Quiet; clear.
,,	26.	N.W.	Quiet; clear.
,,	27.	E., N.E., S.E.	Light; rain.
,,	28.	N.E. to E.	Strong; bitter day.
,,	29.	N.E.	Strong; clear.
,,	30.	N.E. to N.	Hurricane; rain in torrents.
,,	31.	N.	Hurricane; rain in torrents.
Aug.	1.	N.W.	Hurricane; rain.
,,	2.	N.W.	Hurricane; rain.
,,	3.	N.W.	Half gale till 7 P.M.
,,	4.	N.E.	Light; clear.
,,	5.	N.E.	Very light; clear.
,,	6.	N.E. to E.	Still; dripping dense fog all day and night.
,,	7.	N.E.	Light; clear, sunny.
,,	8.	S.E.	Light; clear.
,,	9.	N.E.	Light; rain at night.
,,	10.	S.E.	Light; fine.
,,	11.	S.W.	Still; fine.
,,	12.	S.E.	Light A.M., strong P.M.
,,	13.	S.E.	Light; rain.
,,	14.	N.E.	Still; fine.
,,	15.	S.E.	Light till evening, then strong.
,,	16.	S.	Very light; clear all night, 'most unusual.'

Aug.	17.	N.E.	Light; thunderstorm.
,,	18.	S.	Light; clear.
,,	19.	S.E.	Still; drenching fog.
,,	20.	N.W.	Thunderstorm and deluge of rain.
,,	21.	S.W. to N.E.	Light; dripping fog.
,,	22.	N.E.	Very light; dripping fog.
,,	23.	N.E.	Still; dense fog at night.
,,	24.	N.E.	Still; dripping fog all day and night.
,,	25.	N.E.	Gale in A.M., with driving mist; P.M., heavy rain; evening, thunderstorm.
,,	26.	S.W.	Stiff; clear.
,,	27.	S.W.	Light; clear.
,,	28.	N.	Gale; wild and rainy.
,,	29.	N. to N.E.	Gale and rain.
,,	30.	N.E.	Terrific gale and rain.
,,	31.	N.E.	Gale and rain.
Sept.	1.	N.E.	Gale; less rain.
,,	2.	N.E.	Gale; no rain.
,,	3.	N.E.	Gale died out in early morning; hard frost.
,,	4.	N.E.	Gale again; snow.
,,	5.	N.E.	Gale; snow and rain.
,,	6.	E.	Gale; rain.
,,	7.	N.E.	Blew hard till mid-day, then abated.[1]
,,	8.	W.	Still A.M.; rain and wind P.M.
,,	9.	W. to N.	Still; sun, snow, rain, hail and frost.
,,	10.	S.	Strong; snow.
,,	11.	S. to S.W.	Strong; frost, snow and rain.
,,	12.	S., S.E., E.	Strong; tremendous rain all day.
,,	13.	N.E.	Strong; showery.
,,	14.	N.E.	Very wild; heavy snow.
,,	15.	N., S., W.	Light; clear.
,,	16.	N.E.	Strong wind A.M.; P.M., gale and heavy snow.
,,	17.	N.W.	Calm; fog.

[1] 'With a slight abatement on Monday, the 3rd, we have had ten days of gale.' —*Extract from my Diary.*

INDEX

Printed by T. and A. CONSTABLE, Printers to Her Majesty
at the Edinburgh University Press

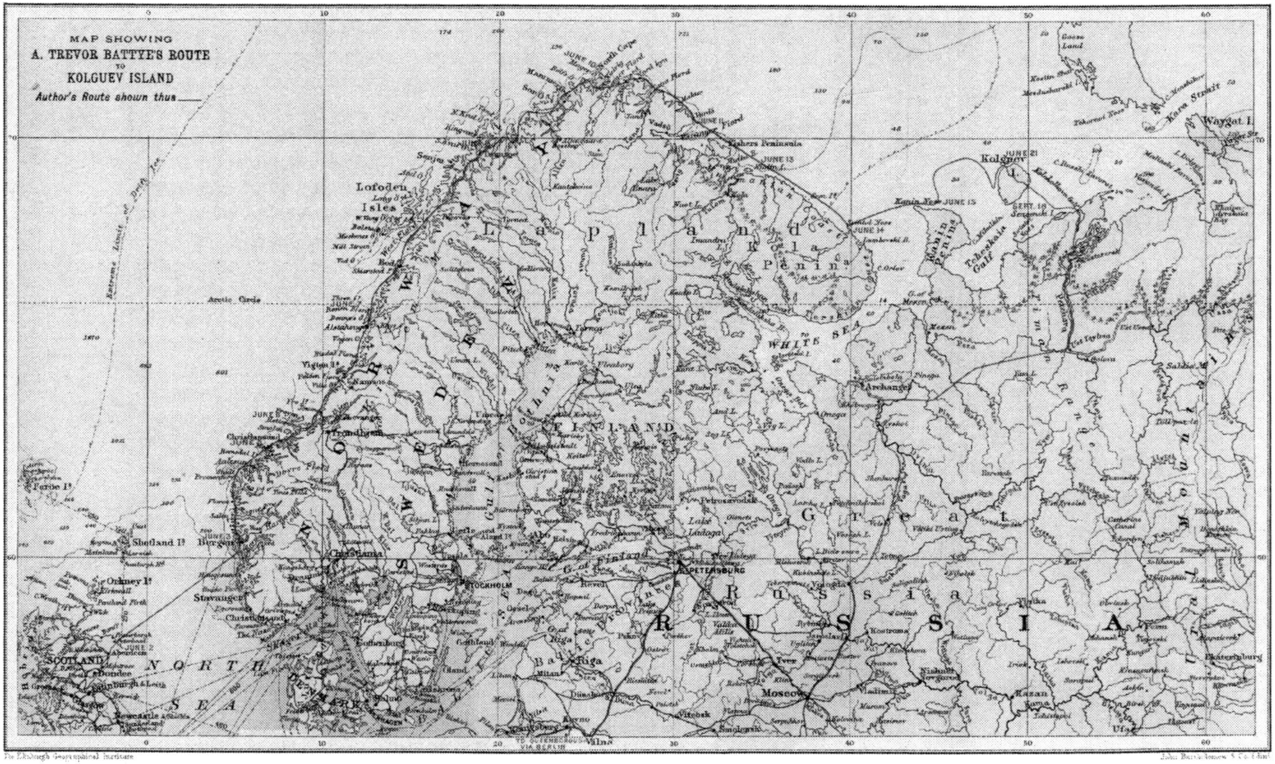

The Edinburgh Geographical Institute
John Bartholomew & Co. Edin.

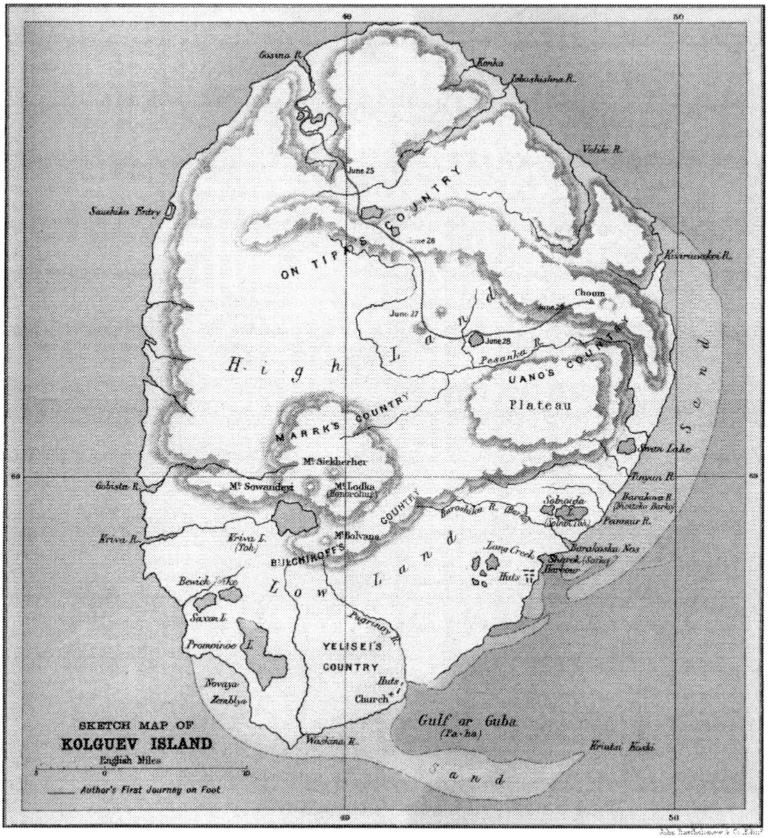

John Bartholomew & Co. Edin.

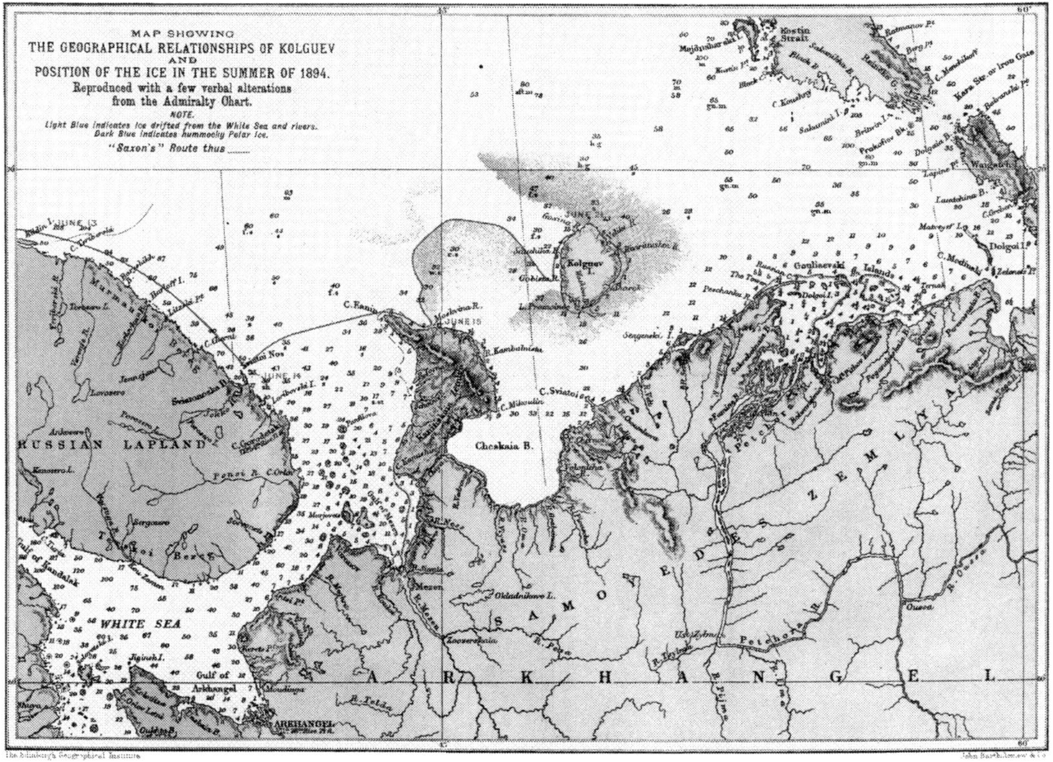
MAP SHOWING
THE GEOGRAPHICAL RELATIONSHIPS OF KOLGUEV
AND
POSITION OF THE ICE IN THE SUMMER OF 1894.
Reproduced with a few verbal alterations
from the Admiralty Chart.
NOTE.
Light Blue indicates ice drifted from the White Sea and rivers.
Dark Blue indicates hummocky Polar ice.
"Saxon's" Route thus
Kostin Strait
Kolguev I.
C. Kanin
C. Sviatoi
Cheskaia B.
RUSSIAN LAPLAND
WHITE SEA
Gulf of Arkhangel
ARKHANGEL
SAMOYEDES
ZEMLYA
ARKHANGEL
Goulaevski Islands
Dolgoi I.
Mezen
R. Petchora
The Edinburgh Geographical Institute
John Bartholomew & Co.

Printed in Dunstable, United Kingdom

91224054R00306